Winnetka-Northfield Public Library
W9-DGC-828

EVIDENCE OF THINGS SEEN

ALSO BY SARAH WEINMAN

Troubled Daughters, Twisted Wives:
Stories From the Trailblazers of Domestic Suspense

Women Crime Writers:
Four Suspense Novels of the 1940s

Women Crime Writers:
Four Suspense Novels of the 1950s

The Real Lolita

Unspeakable Acts: True Tales of
Crime, Murder, Deceit, and Obsession

Scoundrel

EVIDENCE OF THINGS SEEN

TRUE CRIME IN AN ERA OF RECKONING

EDITED BY SARAH WEINMAN

ecco

An Imprint of HarperCollins*Publishers*

HarperCollins books may be purchased for educational, business, or sales promotional use. For information, please email the Special Markets Department at SPsales@harpercollins.com.

FIRST EDITION

Pages 275–76 are an extension of this copyright page.

Designed by Jennifer Chung
Ripped paper © autsawin uttisin/Shutterstock

Library of Congress Cataloging-in-Publication Data has been applied for.

ISBN 978-0-06-332392-6 (Library Edition)

23 24 25 26 27 LBC 5 4 3 2 1

CONTENTS

PART II

THE TRUE CRIME STORIES WE TELL

PART III

SHARDS OF JUSTICE

INTRODUCTION

BY RABIA CHAUDRY

The debate about whether the true crime genre, across all forms of media, does more harm than good in society is long-standing and contentious. For many years now, in the perceived wake of the true crime boom, there has been an equal and opposite reaction against it.

The criticism stems from different corners, enumerating several ways that true crime media may harm us. On one hand, popular culture critics like Laura Bogart, who penned a piece unequivocally titled "Why Our True Crime Obsession Is Bad for Society," worry about the glorification and profiteering from violence and violent offenders. She observes that "Poorly executed crime stories . . . equate brutality with profundity" and correctly points out that monsters like Ted Bundy and Jeffrey Dahmer have left permanent marks in pop culture while their victims have been forgotten.

On another hand, there is a real concern that our collective mental health is being damaged by true crime. This worry is not just raised by online commentators and members of the media, but also mental health professionals. In 2021, citing the rise of true crime podcasts, books, and TV shows, the Cleveland Clinic posed the question "Is Your Love of True Crime Impacting Your Mental Health?" They answer this question with a "maybe." In the right (or wrong) circumstances, true crime can increase anxiety, impact social skills, disturb sleep, and even lead to hyperventilation and heart palpitations. Stories about murder, rape—all sorts of unspeakable violence—can lead to paralyzing paranoia and fear, especially in women, they argue. These stories create distrust and fear of strangers, preventing some from even engaging in small talk with neighbors, certain that anyone, anywhere could be the next John Wayne Gacy or Aileen Wuornos. No small talk, say these critics, means less connectivity and less personal happiness. Pay attention to how your body reacts to true crime, they advise, and take a break when needed.

Both of these arguments have a shared underlying assumption: that true crime has only recently become a national—nay, international—"obsession." And it is this explosion of true crime that is at the root of our problems. We are surrounded by it, unable to escape.

This, however, is a false premise. Public fascination with true crime is as old as human history. Crime and criminals are centered in the greatest stories ever told. From scripture to Shakespeare, there is no great epic without crime at the heart of it. Violent crimes in particular expose what happens when the darkest aspects of human nature—envy, greed, lust, pride—go unchecked. They serve as lessons for us, mirrors held up to our own internal urges, and assurances that justice, divine or otherwise, is never far behind.

Long before Ted Bundy became an iconic true crime subject, Jack the Ripper occupied the thoughts and nightmares of many millions (and still does). Who among us didn't grow up singing a children's ditty about Lizzie Borden murdering her parents with an axe, over a century after it allegedly happened? Sir Arthur Conan Doyle, Agatha Christie, and the authors of thousands of sixteenth-century crime reports, nineteenth-century penny dreadfuls, and the Hardy Boys and Nancy Drew series recognized and banked on the ever-present public interest in stories about crime. And what is the nightly news but a brief review of the day's local true crime? All that has changed in the past few decades is that newer forms of media give us faster and easier accessibility, oftentimes with near-instant reporting, and greater global reach. We can stream on-demand documentaries and podcasts not just about local crime stories, but also about crimes committed in India, Australia, Ireland, Israel, without ever leaving the comfort of our couches.

—

So no, the obsession isn't new, and neither is the societal preoccupation with notorious criminals. It's just easier to feed. And even though the average person may have access to more true crime media than ever, it is also a misperception that it has become inescapable—after all, according to Pacific Content, in 2022 only 17 percent of the top podcasts on the charts were true crime shows, and Edison Research reported that in the same year, true crime ranked fourth in the most popular podcast genres. In other words, it only feels like you're surrounded by it if you're immersed inside a true crime bubble by your own choice.

As with true crime's impact on our mental and emotional health, cautions against the genre tend not to be backed by empirical evidence. More study is needed to determine what effect, if any, consumption of true crime stories may have on our psychological well-being. Interestingly, there is existing data—some of which is captured in this very anthology—that suggests the true crime genre is beneficial to society. This collection before you is proof positive that without this genre, advocates like me would be at a loss, struggling to tell the stories of the neglected victims of crime and victims of the criminal justice system.

I was in law school in 1999 when my younger brother's best friend, Adnan Syed, was convicted of the murder of his classmate Hae Min Lee. Adnan was just seventeen years old at the time of his arrest and maintained his innocence from the day the police hauled him in handcuffs from his bed. I believed in his innocence and hoped in vain for fifteen years that appellate courts would give Adnan justice, before I finally turned to media—true crime media. The podcast *Serial* brought Adnan's case much-needed attention, awareness, funding, and support, and made his story an international phenomenon—but didn't exonerate him. So I wrote a book about the case, produced an HBO series about it, and launched my own podcast, *Undisclosed,* with attorneys Susan Simpson and Colin Miller, which first examined Adnan's case and then went on to investigate and report on twenty-four other innocence cases.

Half of the defendants in those cases, including Adnan, are now home thanks to our work, and it was true crime media that made all the difference. Our investigations and reporting uncovered new evidence, including witnesses who had never been spoken to and leads that were never considered before. Through *Undisclosed,* and many other shows like it, we were able to educate an audience of hundreds of millions about how the criminal justice system really works. In the past decade, Americans have become more knowledgeable than ever about police brutality, prosecutorial misconduct, junk forensic science, institutional racism, the phenomenon of false confessions, the impact of cash bail on the poor, and the prison industrial complex—and nearly all of this education can be attributed to true crime media.

And so we have this collection before us, a group of true crime stories that will intrigue but also enrage, fascinate but also educate. We begin with a look at what this nation reckons with, or rather should be reckoning with, through stories everyone should know: the indifference with which violence against Black girls and women is met, a crisis of thousands of missing and murdered Indigenous women across the nation, the lynching of a young Black man sitting unsolved for decades due to police inattention, the untold tales of physical and sexual abuse common to over 90 percent of incarcerated women, the rise in murderous hate crimes against Asian Americans.

These are the stories that usually stay hidden, the ones that have only recently begun to emerge after centuries of true crime focused on the success and power of law enforcement, on the honor and respectability of the system and all who serve in it. These are the stories that have been buried beneath an illusion of American exceptionalism, that are still fighting to be told in an era when teaching the truth about our national history of racism and racial injustice is being outlawed state by state. We rarely hear about these victims, and even more rarely hear directly *from* them, because, as the next section illustrates, *who* tells the story always determines what story gets told.

Who owns the face, name, life of the subject at the center of a true crime story, who decides who the victims, the perpetrators, and the heroes are, what figures are celebrated as trusted sources on crime and justice, and who gets to dole out justice? These are the questions interrogated in the next category of pieces in this anthology, tied together by a common thread—the shaping of public perception, for decades, by unreliable narrators. We have lived through the era of stories about the glorified, duty-bound, gut-guided detective and the victims who had some responsibility for their own demise. We are currently living through a new iteration of American vigilantism, in which gullible online hordes move in swarms to solve mysteries and mete out justice, often in response to information that is at best questionable and at worst outright fiction designed to manipulate the social media user. These pieces challenge us to look beyond the words on the page and question—or at least consider—the source, to acknowledge how our own perceptions and biases have been impacted by dubious narrators, and to be aware of whose voice is centered in media.

Finally, the anthology takes us to the changing face of justice today, to those who are seeking ways to take back their power and searching for redemption. From a program that brings the victims of sexual assault face-to-face with the perpetrators of it to a radio station giving incarcerated men on death row one last chance to reclaim their humanity, these stories take us beyond the initial reporting. What happens in the years after the convictions, after the violence? How do people in true crime stories respond to the trauma they've been dealt? This selection will change how you think about justice, redemption, personal power, and maybe even forgiveness.

Certain criticisms of true crime may prove to be valid, but on the flip side, true crime media has become the most powerful advocacy tool available to bring us the stories that have the potential to change hearts, minds, and systems. It is this very media that challenges the deeply entrenched influence of state actors over the public narrative around crime. There are too many among us who became swept up in the tough-on-crime years, who nodded approvingly at three-strikes-you're-out, who leaned into our televisions, eyes wide, as we heard about superpredators for the first time. Popular true crime has served a vital function in the fight against state power, as a tool of protest and empowerment, and a way of finally adding to the public narrative from the margins and the shadows. As an advocate for the incarcerated innocent and criminal justice reform, I have found nothing more effective than popular true crime media to both educate myself and others about systemic injustices, and challenge our own deeply held misperceptions.

This anthology itself is evidence of the true crime genre's potential to do good, to deliver justice in both large and small measures, by offering a small window into stories we've never heard before, from people society never thought mattered before. And there is none better than Sarah Weinman, with her deep experience thinking and writing about crime, to pull these pieces together as a guide for us all to connect the dots, and do better.

EDITOR'S NOTE

The most powerful piece of true crime–related art I'd seen in years was tucked away in a difficult-to-access corner of a downtown New York City museum. This was not what I expected at the spring 2022 New Museum retrospective for the artist Faith Ringgold, a formidable educator and activist best known for showstopper quilts that present visceral juxtapositions of major facets of Black American history.

The quilts were, justifiably, worth the museum visit. But my attention, at the time and since, kept returning to that corner, near a stairwell connecting the museum's third and fourth floors, where the *Atlanta Children* and the paired *Save Our Children in Atlanta* and *The Screaming Woman* sculptures had been secreted away.

Atlanta Children, on the right, was structured as a chess board, but these were no typical pieces. Each depicted a Black child in distress or pain, or dead. On the left was a female sculpture (the aforementioned *Screaming Woman*) clad in a green dress, a button adorning her right lapel, holding a poster Ringgold had created listing the names of twenty-eight boys, girls, and young men. "This is a commemoration to all those wantonly slain in the dawning of life," Ringgold wrote. "Make it impossible for the sins of hate and indifference to persist in America. Stop child murders!"

These words and images resonated as viscerally with me in 2022 as they must have with every person who has viewed them since their creation in 1981. Ringgold was making art out of the Atlanta child murders, one of the most confounding serial crimes in American memory, a story that resists the usual constraints of true crime storytelling—and yet inspired one of the foundational texts of this ongoing true crime moment, whose title and trajectory are in turn the inspiration for this anthology.

—

BLACK BOYS AND GIRLS WERE DISAPPEARING AND TURNING UP MURDERED IN and around Atlanta. Between 1979 and 1981, as answers proved frustratingly elusive in tandem with the rising body count, parents of the murdered children demanded justice, because it was becoming all too clear there would be none. The subsequent killings of two men, Nathaniel Cater and Jimmy Ray Payne, both in their twenties, led to the arrest of twenty-three-year-old Wayne Williams, tied to their deaths by considerable circumstantial physical evidence. But even though police immediately named Williams as the prime suspect in the child murders—murders which appeared to stop with his conviction for the killings of Cater and Payne in 1982—he was never officially charged with those crimes. Larger issues grew more prominent in the intervening decades: Why hadn't Atlanta police taken the child murders more seriously, and earlier? How had systemic inequalities, from asymmetrical economic status to homelessness and, above all, racism, influenced the trajectory, and the mistakes, of the investigation?

More than forty years later, after countless treatments in books, podcasts, documentaries, and scripted television series, the Atlanta child murders remain a cipher among criminal cases. The lingering lack of total resolution highlights how deeply the system failed the murder victims and their families. This is not, however, because of what remains unknown: it is because of what is known, is evident in plain sight, and still denied wholesale.

No wonder the celebrated writer James Baldwin felt called to explore the Atlanta child murders, first in a long essay for *Playboy,* and then in *The Evidence of Things Not Seen,* published in 1985, two years before his death. Baldwin was two decades removed from the height of his celebrity, when *The Fire Next Time* (1963) had made him the philosopher-king of the civil rights movement, whose work was supposed to bring order out of mounting chaos, and who was unfairly expected to alleviate white guilt and offer them hope as a salve.

But by the early 1980s, America had soured on Baldwin. Readers viewed his later novels and nonfiction as downers, as speaking truths that were no longer fashionable, no longer palatable, no longer a conduit for idealism. As Hilton Als wrote in 1998, "Baldwin's fastidious thought process and his baroque sentences suddenly seemed hopelessly outdated, at once self-aggrandizing and ingratiating." Backlashes against progress, and the rise of Reaganite Republican politics, prevailed. Baldwin still spoke in his own tongue, still called out the essential disparities. But younger generations found greater literary kinship with the works of Toni Morrison, Alice Walker, and Toni Cade Bambara. Searching for different truths blinded people to what was right in front of them: that Baldwin still had the fire, still was nowhere close to the end of the line.

This change in attitude about Baldwin may explain why the

initial critical response to *The Evidence of Things Not Seen* was one of widespread bafflement. Baldwin had no interest in adhering to a typical true crime narrative, or even traditional narratives at all. His lambasting of the Atlanta Police Department and of the city's governmental bodies was a song on repeated refrain, but most people just wanted to turn the music off. His reporting was introspective: while he did visit crime scenes and bore witness to the loved ones left behind, Baldwin ultimately concluded that he could not impose himself upon the parents of the murdered children after they had already suffered such grievous and continuing losses.

What baffled people then makes far more sense now in a society loosed from anything resembling order and of a consensus reality. *The Evidence of Things Not Seen* has rightly grown in stature over time, reconsidered as a forerunner for other important works that showed how crime is woven into the fabric of society, how the legal system is built to fail millions of the marginalized, and how prioritizing collective voices can supersede traditional narratives about perpetrators.

As Baldwin wrote in the book's preface, the Atlanta child murders ultimately created a campaign of sustained terror, and because the totality of that emotion is so great, "terror cannot be remembered. One blots it out." It is not the terror of death, but rather "the terror of being destroyed." That palpable sense of fear permeates every interaction Baldwin has as he grapples, once more, with "having once been a Black child in a white country."

It is not the evidence of what is unseen, but rather what society still refuses to see. When Lady Justice is willfully blind, how damaging are the costs and how irreparable is the harm? As both Faith Ringgold and James Baldwin knew too well, crime has always been a catalyst for greater pain, and no part of the pursuit of justice could alleviate it.

—

THE PAST FEW YEARS HAVE BORNE PERPETUAL WITNESS TO SEISMIC CHANGES that continue to rock the globe: the COVID-19 pandemic; protests against racial injustice, and then a vicious backlash; right-wing extremism and the cancer that is white supremacy; increasingly oscillating temperatures thanks to climate change; the obliteration of reproductive and LGBTQ+ rights; and the distortion of reality put in sharp focus by the insurrection at the Capitol on January 6, 2021.

Most of these major events were shocking, but surprises they were not. They were evidence of what was visible to anyone bothering to pay attention. Denial, however, is a more potent aphrodisiac than looking reality squarely in the eye. It is far easier to cast the perpetrator of a mass shooting, for example, as a lone wolf in the throes of mental derangement rather than a cog in the spinning wheel of more cohesive and reprehensible ideologies firmly rooted in bigotry and conspiracy theory.

True crime cannot be divorced from society because crime is a permanent reflection and culmination of what ails society. And while collective interest in true crime has only grown since the first season of *Serial* in 2014, so too has it morphed into something larger and more troubling, reflecting the acceleration of what we cannot look away from.

Evidence of Things Seen—of course the title is an homage to Baldwin, the Jeremiah of the latter half of the twentieth century—is divided into three parts. "What We Reckon With" examines events of the past few years, as well as the precipitating factors that catalyzed those events. Racial injustices past and present are examined by Pulitzer Prize winner Wesley Lowery in his searching portrait of a 1980s-era lynching, and by Samantha Schuyler, chronicling the brief life and murder of Black Lives Matter activist Toyin Salau.

Canadian writer Brandi Morin delves into the plight of missing and murdered Indigenous women and girls, specifically in California but universally applicable to North America, while Justine van der Leun investigates the continuing failure to treat as victims rather than perpetrators those who endure intimate partner violence and kill their abusers. White-collar crime, and the federal government's steadfast refusal to punish those who engage in it, gets the full treatment by Michael Hobbes, and Atlanta merits the cruel spotlight once more through May Jeong's powerful account of the city's 2021 spa shootings, its effect on Asian communities, and the larger history of exclusion and xenophobia.

"The True Crime Stories We Tell" gives space to critical examinations of the genre and those who helped shape it. Amanda Knox takes back her own narrative and voice in unforgettable fashion, while Diana Moskovitz and Lara Bazelon convey the importance, complications, and damage done by the work of Miami police reporter and author Edna Buchanan, and Baltimore journalist and television showrunner David Simon. The ever-growing interdependence between true crime and those who consume it, and what happens when amateur participation becomes something more sinister, gets a full airing by RF Jurjevics.

The final section, "Shards of Justice," offers some paths forward, both for our deeply fissured legal system and for the true crime genre itself. Amelia Schonbek examines a case of restorative justice with unbounded empathy, while Keri Blakinger, one of the finest criminal justice reporters working today, gives an inside look at death row prisoners finding solace and comfort in a radio station that's specifically targeted to them. Sophie Haigney's moving letter to the child of a victim of gun violence, whose death she witnessed, refracts and upends expectations, while Mallika Rao finds greater meaning, even hope, in an unfathomable murder story.

All fourteen of these pieces, as well as Rabia Chaudry's incisive introduction, reflect true crime's shift from providing answers to asking more questions. As a whole, this anthology is a testament to the discomfort we live in, and must continue to reckon with, in order to hold the true crime genre to higher ethical standards and goals. It is our duty to take the evidence of what we see, tell the truth, and strive for better.

PART I

WHAT WE RECKON WITH

A BRUTAL LYNCHING.
AN INDIFFERENT POLICE FORCE.
A 34-YEAR WAIT FOR JUSTICE.

BY WESLEY LOWERY

In the final weeks of her life, Viola Coggins-Dorsey saw into the future.

The 76-year-old had been sick for years, after her diabetes led to kidney failure. By February 2016 she'd all but stopped eating and been admitted to Emory University Hospital. One night, as her daughter Telisa cued up the patient's favorite song—a twangy gospel track called "Cooling Water"—Coggins-Dorsey made a stunning proclamation.

"They found out who killed Tim," she declared. "What, Mama?" Telisa replied, convinced she'd misheard.

"They found out who killed Tim," Coggins-Dorsey repeated insistently. "I ain't gonna be here for it, but they're gonna get who killed Tim."

For three decades, the October 1983 murder of 23-year-old Timothy Coggins, the fourth of Coggins-Dorsey's eight children, had haunted not just his family but all of Spalding, this rural farming county 45 minutes south of Atlanta. Coggins's mutilated body—stabbed dozens of times, with an "X" like the Confederate battle flag carved into his abdomen—was found in Sunny Side, a poor white part of the county, beneath a massive oak known colloquially as "the Hanging Tree." But investigation into his slaying had gone nowhere, effectively abandoned by the sheriff's department after just two weeks. The Coggins family had long ago given up any hope of closure, and at this point rarely discussed the particulars of the case. Her sick mother, Telisa reasoned, was just talking out of her mind.

Younger than Tim by two years, Telisa was the sibling with whom he was closest. He'd taught her to ride a bicycle, and to navigate her way home from the grocery store on her own. When Telisa gave birth to her first child at 18, Tim was the first to burst into the room to congratulate her. Her brother was funny and outgoing, Telisa told me. He loved to party and would stay out late into the night with old friends or new ones he'd picked up over the course of an evening. Tim was a man with an irresistible smile who had never met a stranger.

She'd been with her brother at the Peoples Choice club—a brick building tucked around the bend of a quiet country road, painted black and tan with an inviting red sign—on the night he disappeared. Back then, on the Black side of Griffin, the largest city in Spalding, it was the place to be on a Friday night. The club had a fully stocked bar and hot barbecue for sale. A tightly packed jumble of bodies filled the room, drawn in by a steady stream of Aretha Franklin and Marvin Gaye and, in the fall of 1983, a lot of Michael Jackson. The dance floor brought out the depths of Tim's charm. He could typically be found at the center of the action, stealing the show. And in recent weeks he'd been seen swaying with a young white woman—a scene that stood out among the almost exclusively Black club goers.

Even in the 1980s, interracial dating was frowned upon in Spalding, where a local Klan chapter still held regular rallies and parades. Carrying on with a white woman might be fine for a Black man up in Atlanta, but change comes slower down here. Tim, at least one family friend had warned him, was flirting with danger.

As Telisa made her way to the club's bathroom that night, she overheard people saying that there were white men outside asking for Tim. Moments later came the last time she'd see her brother alive, as he followed one of those men outside.

No one even realized that Tim had gone missing afterward. It was typical for him to disappear for a few days at a time. He knew everyone around town, so the safe assumption was that he was crashing on someone's couch. Two days passed before sheriff's deputies showed up in the neighborhood holding out gruesome photographs and asking if anyone recognized the dead man who was in them. Telisa Coggins insisted that she did not. She didn't want to admit what she'd known immediately: It was Tim.

The sheriff's department conducted a cursory investigation of the murder and assured the Coggins family that they would get to the bottom of the crime. But before long, the months had stretched to years, and then into decades.

"I think they always knew who did it. But because it was a white man who killed a Black man, they didn't care. They never really tried," Telisa told me recently. The family soon began receiving threats: a bloody T-shirt left in the school bus their stepfather drove each morning, a brick through the living room window with a note warning, "You're next," a decapitated dog placed in the hallway of their home. "We knew from the beginning that he had been killed because he was Black."

Neither the crime nor the threats were ever solved. And no one in the Coggins family had any expectation they ever would be. But then, in 2017, about one year after Coggins-Dorsey made her deathbed prediction, the district attorney's office called.

The sheriff's department now knew who killed Tim, the voice on the other end of the phone told them. The investigation had been reopened. After all this time, investigators vowed they would deliver justice.

—

IN THE RURAL REGIONS OF THE STATE, MAJOR CRIMES ARE HANDLED BY THE Georgia Bureau of Investigation, which has 350 investigators. Every six months, the agency cycles its unsolved cases, even those that are decades old, to new investigators in the hope that fresh eyes spot something—which is how, in 2016, the long-abandoned Timothy Coggins case file landed on the desk of Special Agent Jared Coleman, a studious young investigator in his second year at the agency.

He was immediately struck not by what was included in the relatively thin file but, rather, by what wasn't. Police interviews had pointed toward two men: Frankie Gebhardt and Bill Moore Sr., white brothers-in-law who lived in the trailer park near where Coggins's body was found. But although police did interview Gebhardt, neither man had faced much scrutiny. Gebhardt's alibi presented some obvious holes, yet detectives had never followed up. And,

Coleman said, it didn't appear Moore had ever been interviewed at all.

And in recent years an inmate named Christopher Vaughn, who as a 10-year-old had been among the group of squirrel hunters who discovered Coggins's body in 1983, had written to investigators to say that Gebhardt had admitted to him multiple times over the decades that he had committed the killing and thrown the murder weapon down the well behind his trailer. The first confession had been in a comment Vaughn overheard at a party, not long after the killing. But on later occasions, as Vaughn grew older, Gebhardt would bring up the killing, Vaughn claimed. To Coleman's shock, those leads had prompted almost no follow-up.

"The case really hadn't been fully dived into since 1983," said Coleman, who was struck by just how close the trailer park where his major suspects lived was to the out-of-the-way place Coggins's body had been found. When he finally interviewed Moore, Coleman found him evasive: The man claimed to have never heard of the gruesome murder that had for years been the talk of the town. "I could tell that he wasn't telling the truth," said Coleman.

Coleman approached the recently elected local sheriff, Darrell Dix, a tough-talking boulder of a man eager to strengthen the department's relationship with the county's Black residents. Dix had been troubled, not long after his election, to discover documents suggesting that at the time Coggins was killed, a number of the department's deputies were active members of the local Klan—raising questions in his mind as to whether the failure to solve Coggins's murder was not just a failure of police work but, rather, deliberate complicity. Dix assigned a deputy to work with Coleman, and encouraged the men to figure out once and for all who had killed Coggins.

They revisited the crime scene, an open field tucked between cornfields in the shadow of a towering power line, and compared what they saw to the photos taken at the time of Coggins's murder. He'd been stabbed dozens of times and dragged behind a truck along the power line. It was clear his killing was carried out with personal animus: Coggins had been tortured and left to die.

"The death of Mr. Coggins," Coleman would later tell me, "was very clearly a lynching."

According to a recent estimate by the Equal Justice Initiative, thousands of Black Americans were lynched in the decades between the Civil War and World War II—with Georgia second only to Mississippi in raw numbers killed. These brutal extrajudicial killings, often staged as wicked public spectacles, took place across the country but were especially pronounced in the south, where white citizens shared both fear and resentment toward their now-emancipated Black neighbors. With American slavery vanquished, white southerners were now determined to wield indiscriminate terror to maintain a societal system of white supremacy. A quarter of southern lynchings, the EJI study found, were fueled by obsessive revulsion at the concept of sexual contact between Black men and white women.

"The miscegenation laws of the South . . . leave the white man free to seduce all the colored girls he can, but it is death to the colored man who yields to the force and advances of a similar attraction in white women," the journalist Ida B. Wells wrote in her 1892 pamphlet *Southern Horrors*. "White men lynch the offending Afro-American, not because he is a despoiler of virtue, but because he succumbs to the smiles of white women."

Eventually, these lynching spectacles died out. Yet white vigilante violence never fully disappeared, nor did the promise of impunity for its perpetrators. The message, codified by the shameless crimes of generations of white southerners: White men are free to steal the lives of Black men—especially those who've pursued a white woman.

As Coleman looked more closely at Coggins's death, he began to see a familiar story.

—

FRANKIE GEBHARDT WAS ALREADY IN CUSTODY AT THE SPALDING COUNTY DE-tention Center on an unrelated sexual assault charge when, in April 2017, investigators came to question him about the 34-year-old lynching.

"I don't know a damn thing about that," the 59-year-old inmate insisted.

Gebhardt said he didn't remember hearing about the murder and definitely didn't have anything to do with it. He didn't remember ever bragging to anyone about having committed it either. But then again, after 23 years spent as a drunk, Gebhardt conceded, there was a lot he didn't remember. When investigators asked about the rumors that he'd thrown the murder weapon down his well, Gebhardt quipped back, "Well, y'all come out there and dig my damn well up." When, toward the end of the interview, Coleman showed him the photo, Gebhardt erupted. "I ain't never seen that picture," he exclaimed. "I ain't never seen that nigger."

Gebhardt had spent his entire life living at or near Carey's Mobile Home. Having dropped out of school after sixth grade, he supported himself by working shifts logging timber. On the weekends he'd host wild, debaucherous parties featuring beer and pills and shrooms and, at least one time, the drunken butchering of a cow on the kitchen floor of one of the trailers. For years he'd been inseparable from his brother-in-law, Bill Moore, who like Gebhardt had a reputation for violence. They were known as "frequent fliers" at the local courthouse.

"Just a regular guy who was brought up kind of rough," explained Larkin Lee, Gebhardt's attorney, who acknowledged his client was "no stranger to drinking and fighting" and had a "propensity" toward racial slurs. Still, Lee said Gebhardt has always denied to him that he had any involvement in Coggins's death. "I think a lot of people have heard it over the years. I'm not sure that Frankie has ever said it," Lee told me. "It's one of those things where it's a rumor, and 30 years later people swear that they've heard it directly from him."

But as Coleman worked the case, he encountered person after person who insisted that Gebhardt had, in fact, bragged about the crime. They generally claimed that Gebhardt had discovered that Coggins had been sleeping with his "old lady," a white woman who went by "Mickey," and that Coggins had previously ripped off Gebhardt in a drug deal. (Coleman identified "Mickey" as Ruth Elizabeth Gay, who left the state permanently after Coggins's murder and died in 2010.) And so he and Moore picked up Coggins from the nightclub, took him to the Hanging Tree, stabbed him, dragged him from the back of their truck, and left him for dead.

The reports of Gebhardt's confessions varied. An inmate who'd been friends with Gebhardt and Moore years earlier said the men had boasted of how they'd dragged Coggins. The first words from one longtime resident of the trailer park, upon hearing why investigators were at his door, were: "Frankie Gebhardt killed that boy." An ex-girlfriend told Coleman that Gebhardt would beat her while threatening, "If you keep on, you are going to wind up like that nigger in the ditch." A man whose mother had once dated Gebhardt recalled both him and Moore admitting that they'd committed the murder, with the latter drunkenly lamenting the passing of "the old days" of "killing Black people for no reason." For over 30 years, there had been plenty of witnesses. But no one had bothered to seek them out.

Even after Gebhardt became aware that investigators were zeroing in on him, he kept telling people about the crime. At one point, police executing a search warrant seized 60 knives from his trailer. Not long afterward, a new inmate came forward to speak to investigators. He told them Gebhardt had recently confessed to having stabbed Coggins, bragging that investigators had just seized 60 of his knives but that he had disposed of any evidence years ago.

Before long, arrest warrants had been issued for both Gebhardt and Moore, whose families and friends insisted the men were being railroaded by an overeager sheriff's office. "We have no knowledge of this Timothy Coggins case," insisted Brandy Abercrombie, 41, Moore's daughter and Gebhardt's niece. "I've never heard of this [case] in my entire life."

It had taken decades, but there were finally charges in the death of Timothy Coggins. Investigators had their suspects, but with murder trials looming, they were still short on something crucial: evidence.

—

IN HIS FIRST MEETING AT THE PROSECUTOR'S OFFICE, COLEMAN CALMLY EXplained how he'd stumbled across a decades-old cold case he believed they could solve. There were some problems, he conceded. Almost everything from the original crime scene had disappeared: the soil samples and tire tracks, the DNA collected from the slain man's body, the wooden club possibly used to beat Coggins, the empty Jack Daniel's bottle discarded near the scene, the hair samples collected from the victim's sweater and jeans—all of it lost during the years the case sat cold.

The lack of evidence flustered Marie Broder, a sharp-witted 34-year-old prosecutor who'd been the protégé of Layla Zon, a prosecutor turned judge in the nearby Alcovy Judicial Circuit. Zon had taught her how to be aggressive and firm, but not so firm that she'd be written off by her colleagues, the judge, or, most crucially, jurors.

Broder zeroed in on finding the piece of evidence that could cement the case: the knife used to kill Coggins. Investigators knew that if the murder weapon still existed, it was most likely at the bottom of Gebhardt's well. But that presented a problem: The well was too close to the trailer, impossible to excavate without destroying the house.

"We need to get in this damn well," Broder insisted to Coleman during a phone call one night.

"I'll find a way to get into it," Coleman assured.

Soon enough, they'd located a hydrovac company in Atlanta that could blast water into the well and then vacuum up any loose debris without destroying the trailer. Before long, they were sucking up years of dirt and trash.

When they emptied out the vacuum tank, they discovered a bounty of evidence: a pair of Adidas shoes like the ones Coggins's family said he was likely wearing on the night he vanished, a white T-shirt that appeared to be torn by multiple stab marks, and, most crucially, an old broken knife.

They had their evidence. Now it was time to prepare for trial.

—

GEBHARDT'S DEFENSE, ANCHORED BY LEE—CONSIDERED BY PROSECUTORS TO be the most skilled defense attorney in the circuit—would be simple. Sure, Gebhardt said racist things. But he was no less credible than most of the felons and inmates and drug users testifying against him.

All the defense needed was for a single juror to decide that all of these witnesses were lying and hear a reasonable doubt of Gebhardt's guilt (Moore decided to forgo a trial).

"Ain't nothing guaranteed," said Telisa Coggins. "As a Black woman, a Black human being, living in a racist town, you never know what is going to happen."

Broder was racked with nervousness as she prepared her opening statement. She lost 10 pounds over the course of the weeklong trial. In moments of doubt, she'd look back at the Coggins family, who packed the court gallery each day, to renew her determination. She was so dialed in that, for the first time in her career, she forgot to overthink. Broder didn't calibrate her words or tone, allowing the passion of her rhetoric to match the heinousness of the crime.

"It deserved fire and passion. I wanted those jurors mad about what happened to Tim Coggins," Broder would later tell me. "I wanted them rocking back on their heels."

Prosecutors called more than a dozen witnesses: the medical examiner who detailed the various wounds to Coggins's body, friends and family members who testified that Coggins had been dating a blue-eyed brunette, and seven people—residents of the trailer park and inmates at various correctional facilities, including some who checked both boxes—who said Gebhardt had admitted to them that he had committed the murder.

"He would smile when he spoke of it," testified Charlie Sturgil, a current inmate who had grown up in the trailer park. Patrick Douglas, who worked as the prison barber and was a member of the Aryan Brotherhood, with white supremacist tattoos up and down his arms, told the court that Gebhardt had approached him, claimed to be a member of the Ku Klux Klan, and then confessed to killing Coggins. "Seemed he was excited when he done it," Douglas testified.

The defense called just two witnesses, former GBI agents who had previously worked on the case, in the hope that their testimony would help demonstrate that prosecutors had no stronger evidence now than their predecessors did years earlier. Gebhardt declined to take the stand in his own defense.

"It's a made-up story. It's a reasonable doubt, because it's a made-up story," Lee declared during his closing arguments. Investigators had, for decades, failed to properly investigate Coggins's death and retain evidence, and now, Lee said, they were using a parade of inmate witnesses in an attempt to wrongfully convict his client. "And that's what you get when you bring in people who are dressed in street clothes but have left their striped jumpsuit right behind that door over there, because that's what they're wearing normally.

"It's just trash," he continued. "That's what those witnesses amount to. That's what all your jailhouse witnesses amount to is just trash. The same thing that was found in the well."

But in the end, it was Gebhardt's own boasts, stretching from the days after the murder to just weeks before the trial, that convicted him. "We counted 17 times that Mr. Gebhardt admitted to the murder in some kind of way over the years," the jury foreman would later say. Broder fidgeted in her seat as the moment approached, and buried her head in her folded hands as the judge read the verdict: guilty on all five counts. Members of the Coggins family broke into tears. Gebhardt kept his unblinking eyes trained on the judge.

"I'm grateful we were able to bring justice for them," said Coleman, who has since moved to the GBI's Gang Task Force. "Mr. Coggins is not forgotten."

"This case changed me forever," added Broder, who has since been appointed district attorney. "I had never experienced evil purely based on someone else's skin. You really know nothing, and you have to recognize that and say: This happened, it happens. And in order to confront this evil, you cannot shy away from it. You have to confront it head on."

The judge sentenced Gebhardt to life in prison: "Hopefully, sir, you have stabbed your last victim," he declared from the bench. (After Gebhardt's conviction, Moore agreed to plead guilty to manslaughter in exchange for a 20-year sentence.) As the courtroom emptied out, members of the Coggins family found themselves just a few feet from Bill Moore's daughter, Abercrombie, who was overcome with emotion. The two families had talked on occasion throughout the trial, and the Cogginses felt bad for Abercrombie—she'd been just a little girl at the time of the murder, so of course she'd have trouble believing her father and uncle could have committed it.

"I'm sorry this happened to your family," Abercrombie sobbed, as Telisa Coggins wiped the tears from the distraught woman's face.

"Black people have a way—because of all that we've been through, the way we was raised—forgiveness is the first thing that Black people learn," Telisa recently told me, laughing as she remembered her mother and the deathbed prediction. "After all of the stuff that Black people have endured, from slavery up until now, we still are a forgiving people."

ORIGINALLY PUBLISHED IN GQ, JULY 2020

THE SHORT LIFE OF TOYIN SALAU AND A LEGACY STILL AT WORK

BY SAMANTHA SCHUYLER

Dusk was waning on June 14 when her friends got the news: One of the two bodies police had found the night before had been identified. It was Toyin. A 19-year-old student living in Tallahassee, Oluwatoyin Salau had met the group of activists during the George Floyd protests, a moment when young people across the city were flooding the streets to demand justice. The protesters had grown close quickly, bonding over shared traumas and a vision for the world. That is, until Toyin slipped away from a protest one night and, soon after, posted a cryptic and distressing Twitter thread that described a sexual assault. Her friends knew that Toyin had been assaulted in March and that the man was still harassing her. For her to go missing seemed disastrous.

After a week of searching for her, the news seemed like a bad dream. As the community spread the word throughout their network, Ashley Laurent, a 22-year-old student at Florida A&M University, sent out the news to the people on Instagram and Twitter who had been asking her almost daily for updates. Shaken and queasy, Ashley pressed send: "I'm sorry to inform everyone of this, but Toyin is no longer with us."

In the days that followed Oluwatoyin Salau's death, Ashley could only watch as her friend's name became a hashtag. Much to the surprise of the protesters in Tallahassee, her disappearance—just hours after she posted the details of the sexual assault—and death had thundered across the internet. Around the world people painted portraits and planned vigils. Elizabeth Warren, Kamala Harris, and others made public statements expressing their sadness. "I'm furious. I'm heartbroken. You deserved protection," wrote Ari Lennox in an Instagram post. Barry Jenkins, who graduated from Florida State's film school, pointed out that Tallahassee was Florida's capital—and even there, Toyin had not been protected. "I am her and she is me," wrote Gabrielle Union. "I am alive to talk about surviving my rape at 19. She is not." Even Kehlani, one of Toyin's favorite artists, wrote a long tribute.

Alongside the hashtag #JusticeForToyin, Black women pointed out that the injustice of Toyin's murder reflected a wider issue: that Black women and girls experience disproportionate amounts of sexual violence—a material, demonstrable, and exhausting reality that is regularly met with indifference. "The weight of existence and the trauma of unresolved violence that Black women carry?" wrote Folu Akinkuoto on Twitter. "It's just so unfathomably painful and fucked."

Toyin's friends felt dizzied by the paintings and poems flashing across their feeds, a bulletin of grief. But as thousands of people retweeted photos and videos of Toyin, sharing her story so widely that she became a trending topic on Twitter, they knew they were watching their friend become a symbol. For them, Toyin's murder could never be figurative. They knew her laugh, the music that she sang along to, how she would change her hair on a whim. They saw firsthand the lack of value attributed to her life, that even when crowds of people rallied to find her, it wasn't enough to merit the same urgency as a white woman. These young Black women never stopped agitating about Toyin's disappearance: They forced the public and police to pay attention, to prevent their friend from becoming a statistic. Without them, Toyin's name could have been swept to the side, like any of the 64,000 or more Black women and girls who are currently missing in the United States, a statistic only exacerbated by the well-documented disparity that Black women's disappearances are often erased by police and in the press—what Gwen Ifill once nicknamed "Missing White Woman Syndrome."

Notably, Toyin hadn't been found until police launched a search for Victoria Sims, a 74-year-old white woman and longtime community volunteer. Just hours after Sims's family had reported her missing, a fleet of officers were dispatched with search dogs after getting a warrant to track her phone. Within hours the police had located her body. That's how, more than a week after Toyin went missing, search dogs stumbled upon her body, in an adjacent plot of land, covered in leaves.

From the day they last saw Toyin, the young activists knew they couldn't rely on the police or local officials. Instead, the young people who flooded the streets to protest the murder of George Floyd organized searches, they demanded accountability, and they made the world pay attention. They also demonstrated how police, even when not acting punitively, can enact another form of violence: deprioritizing, devaluing, and ultimately not believing threats to Black women. Now, they're the ones honoring her death. Those with access to more resources failed to take Toyin's repeated cries for help seriously; they had failed to use their power to keep her safe. In the weeks leading up to her death, Toyin was fighting for a world that wouldn't have failed her like this one did.

"The community protects the people more than the police does," Danaya Hemphill, a 23-year-old FAMU student who helped lead the searches, told me when we met in Tallahassee, at a house not far from where Toyin was found. Wreathed in the muffled drone of cicadas, she leaned forward on the couch we were sharing, her voice quickening with the determined edge it would take on every time we would talk about Toyin. "Because who was it who showed up for Toyin? The community. Not the police."

—

LIKE MANY OF THE ACTIVISTS IN TALLAHASSEE, DANAYA (WHO GOES BY DANI) and Ashley first met Toyin at the end of May, while the protests for George Floyd were ricocheting around the country, bringing people who never knew each other together. They had met Toyin on the city's first day of protests, a week before she went missing.

Tallahassee is a small city. Its organizing circles were close-knit before the uprisings, and the George Floyd protests drew hundreds of new people into the fold. The city had seen three officer-involved shootings in a span of three months: one in March and two in May. Over that week, the protesters had grown close to each other, forging bonds and friendships that they say have changed their lives. "Tallahassee felt it personally," Octavia Thomas, a 23-year-old graduate of Florida State University and organizer with the Movement 850, a student activist collective, told me. "It was in our backyard."

They were organizing together and bailing each other out of jail. They were marching in the streets, then coming together to talk about what they wanted for the future. Toyin had become a part of this community—especially among the young Black women at the forefront. "That is like your soul sister. That run deep," Dani told me. "It's like: I got your back. At protests, we come together, we leave together . . . Don't let anything happen to them."

On May 29, the first day of protests in honor of George Floyd in the city, hundreds gathered at the capitol building to march to the Tallahassee Police Department headquarters. Flanked on both sides by friends, Ashley was scanning the crowd when she noticed Toyin sitting on the capitol steps, leaning against a pillar. She was crying and holding a sign with Tony McDade's name written in black ink. Two days before, on May 27, McDade, a Black trans man, had been shot and killed by Tallahassee police in his apartment complex. Hours before his death, he had made a Facebook Live recording that recounted how he'd been attacked and beaten by a group of men who he believed were targeting him for his appearance and gender identity.

Ashley broke away to sit beside Toyin. She introduced herself. Gently, she asked what was wrong. "Nobody is saying Tony McDade's name," Toyin said. Ashley listened and rubbed Toyin's back, assuring her that they'd say his name, too. Toyin said she hadn't realized there was going to be a big protest—she lived nearby and had come to the capitol alone to sit with her sign. Ashley and one of the organizers of the protest, Ashleigh Hall, a 22-year-old FAMU student, encouraged her to march with them.

Toyin talked to them about Blackness, about God, about

oppression. She became more and more emotional. When they got to the police department, people began to take turns speaking into a local news station's camera. Ashley nudged her. "You got a voice," she told Toyin. Toyin hesitated. She was shy. Ashley told me she told Toyin, "Come on . . . You want them to know the story of Tony McDade? You should go do it."

Ashley took her by the elbow and brought her to the middle of the crowd. She tapped the reporter for the local TV station on the shoulder. "My good sis would like to speak," she told him. So Toyin did. Toyin spoke about how African immigrants to America—like her own family, who had emigrated from Nigeria right before she was born—needed to be sensitive to African American history. Then she brought up Tony McDade. "We're all brothers and sisters out here, but the fact that I felt his pain is not OK," she said. "It's not OK. They shot him in cold blood." Dani hadn't realized at the time that Tony McDade was a trans man. (TPD and a local television station initially misgendered McDade.) Toyin's insistence on saying his name struck her—she admired the courage it took to hold fellow protesters accountable. Dani decided there that they would be friends.

Jaelyn Guyton, another FAMU student and organizer with the Movement 850, told me Toyin's speech that day changed the protests going forward. "Toyin ensured we focused on Tony McDade," he told me. "Queer lives are often erased by history. It means a lot for her to have done something like that." Ashleigh Hall said that they hadn't planned to say his name, but that Toyin insisted, explaining the particular violence experienced by trans people. "He deserves justice, too," she recalled Toyin saying. "She sparked that fire." Other protesters saw that though she seemed shy, when given the mic she came to life. "She really shined," said Delilah Pierre, an organizer with the Tallahassee Community Action Committee (TCAC), a local grassroots activist group that organized several of the George Floyd protests. "That passion was real. You can always tell. And I could tell immediately that Toyin had this care and this fire." At a protest later that week, Chancellor Crump, a 22-year-old student at Tallahassee Community College and adopted son of Ben Crump, the attorney representing the families of George Floyd and Breonna Taylor, said that Toyin had come up to him when she noticed he was nervous about speaking. He said he had been close to giving up and not saying anything at all, but that she encouraged him. "She told me, 'You have a voice,'" he said. "'Just speak how you feel.'" So he did. His prayer and speech made it into the local paper.

A few days after they first met, Ashley was speaking to a crowd of protesters when she saw Toyin again. "And I was like, 'Oh my God, that's my friend! Come up here!'" She passed Toyin the mic. Later in the march, in a video that has since become recognized across the country, Toyin spoke directly into the microphone, unfurling a speech that brought encouraging whoops from the crowd and nearly 10 million views.

In the video, protesters mill behind her on the steps of the capitol building, fanning themselves in the heat, clapping and nodding occasionally. "Tony McDade was a Black trans man," she tells the camera. "OK? We doing this for him. We doing this for our brothers and our sisters who got shot. We doing this for every Black person. Because at the end of the day, I cannot take my fucking skin color off. I cannot mask this shit, OK? Everywhere I fucking go, I am profiled, whether I like it or not. I'm looked at whether I like it or not."

At this, she takes a step back, gesturing to her face, glistening with sweat, and to her locs, bouncing with each word that she launches at the camera. "Look at my fucking hair. Look at my *skin.* I can't take this shit off. So guess what? I'mma die by my fucking skin."

As she raises her voice to say this, the crowd comes to life, breaking into cheers and applause. "You cannot take my fucking Blackness away from me. My Blackness is not for your *fucking* consumption."

As the cheers erupt around her, she gazes, brow furrowed, into the camera, indifferent to the noise. The protesters vocally supported her. Some may even have understood, at a deeper level, the pain and fury that crackled in her voice. But nobody there knew, not really, what she was feeling in that moment.

—

ON THURSDAY, JUNE 4, AFTER NEARLY A WEEK OF DAILY PROTESTS, ASHLEY woke from a nap to her phone buzzing. It was Toyin. She was upset. She told Ashley that a man who had assaulted her back in March had tried to force himself on her again, in her home. He was gone, but his friend was still there.

Ashley, shaking off a fog of sleep, heard panic in Toyin's voice. She made a plan: When she got off the phone, she would call other women in the activist network, and they would come get Toyin. This was the first Ashley was hearing about the assault or anything about Toyin's history with trauma. She didn't know a lot, but she didn't need to: She could feel Toyin's distress through the phone, and it alarmed her. "You're not going to stay inside that house," Ashley told her.

She picked Toyin up, and brought her to New Life United Methodist, where Ashley had been planning to spend the afternoon making sandwiches for the protests. There, Toyin called a lawyer but declined to write a statement about the man. She came out of the room crying. "She was like, 'I'm not a victim,'" Ashley recalled. "She said if she pressed charges, then he would get killed by the police and become a hashtag. I got so mad."

Toyin's response was rational considering the history of disproportionate police violence against Black people. According to a 2019 study in the *Journal of Urban Health,* Black women often consider fear of an "overzealous law enforcement response" as one of many reasons for not reporting their assault. Overall, only 23 percent of sexual assaults are reported—a distressingly small number that is in part the result of an inherently sexist criminal justice system. Survivors point to concerns of being retraumatized, feelings of self-blame, fear of reprisal, and a lack of perceived "proof" or injury as some of the reasons they do not report—a set of reasons that is only exacerbated for Black women. "This navigation of competing priorities is emblematic of how Black Americans have been taught to manage police dynamics in light of historical discrimination and ongoing instances of police violence," the study reads. "This barrier was not discussed by any White study participants."

Black women deal with overlapping structures of inequality. In addition to navigating a sexist system, the police's historical, discriminatory violence against Black people—indeed, the very reason for the protests that brought the young women together—and the country's history of disproportionate mass incarceration are twin pillars of a prison industrial complex that wreaks unmitigated terror on Black communities. For Toyin and other Black women, the lack of alternative recourse becomes a double bind, one that reflects the reality of negotiating trauma and sexual violence in a world that does not value Black lives.

Instead, Ashley called the church pastor, Latricia Scriven, and asked her to speak with Toyin over the phone. They had never met before. At first, over FaceTime, they spoke vaguely about living through pain, but soon Toyin began to describe her March assault. Then she explained what had happened earlier that day. Scriven remembers her saying emphatically she was not a victim. "I know

I'm fighting for things I've been fighting for, for a very long time," Scriven recalled Toyin telling her. "These are my brothers and sisters in Christ, so I know I just have to forgive them." Scriven listened, then told her that even if she forgave them, that did not mean what she experienced was OK, or that there could not be consequences. They prayed together.

Ashley and Tamra, another FAMU student and activist, took Toyin to her house to help get her things. Meanwhile, Dani began texting people to organize a chain of places Toyin could stay until she found stable housing. They plumbed their network to gather extra clothes, toiletries, and meals. "I saw a young woman in trouble surrounded by friends who wanted to help," Scriven told me. "These groups of students, who are becoming her community, were asking the question, 'What will happen to her if we will not try to help?' And this is the meaning of being a good neighbor."

They called the Tallahassee Police Department and requested a police escort, who arrived when they did. He followed them inside, as they picked up her clothes and belongings, bundling everything in her bedsheet. In a video Ashley took at the time, documenting the interaction with police in case something went wrong, Toyin sobs as she leads them through the house, moaning, "Oh God, oh God almighty." Her friends grabbed a painting she had made, her mannequin she used to practice for cosmetology school, her immaculate white sneakers. Ashley reminded her to get her shea butter.

According to Ashley, when they had called the police to request the escort and report the assault, the department asked Toyin if she were "a victim or hurt." Toyin declined either characterization—as she repeated again and again, she did not consider herself a victim. And the assault had not been technically violent—she wasn't injured. Because of this, the officer had shrugged when they demanded that he take some kind of action. Toyin had just confessed that her abuser had been in this house and had harassed her here—wasn't TPD going to even try to investigate? "There's not enough evidence," they recalled him telling them; the man had already left the house, and they didn't know where he had gone. "There's not much that we can do." (When asked about this, TPD's public information officer told me they could not comment on the case, as it is still open and now under grand jury investigation.)

For Dani and Ashley, the police's apparent indifference showed them how little their concern and distress meant to the people they were told protected them. What they found frustrating, they believe was traumatic for Toyin. Even when she did everything right, seeking help and notifying the police, there wasn't much they could do. Her assaulter could simply walk away. "If my Black ass was to go and get pulled over with a gram of weed, they will pull up six squad cars deep," Dani said. Where was that urgency when a woman says she's being harassed?

Toyin stayed at Dani's place that night. She took a shower and changed into Dani's clothes. She had blisters from wearing Jordans with no socks to the protests every day, so Dani gave her the white Crocs she used at her job as a vet tech. Sitting on Dani's bed, they talked about their shared traumas. Toyin talked about getting kicked out of her childhood home and having to stay at friends' houses through high school. Dani told her about her dad's and brother's deaths. They talked about boys and crushes. They talked about Nigeria, where Toyin's parents had emigrated from shortly before she was born. Dani, surprised, told her the names of her dog and her cat: both Yoruba words. Soon, they went to bed; Toyin slept on the left side.

The next day, Friday morning, Dani, Ashley, and Toyin drove

to the county courthouse and collected any paperwork they saw relating to sexual violence, including a form to file a restraining order. They went to Ashley's apartment and deep conditioned Toyin's hair. Dani oiled her scalp. It was like any gathering they had before a protest: communal and electric with anticipation—the young women rushed between the bathroom and the kitchen to the bedroom, dancing to conversation that was buoyed by peals of laughter. Ashley's mom FaceTimed her and said hi to Ashley's new friend—they all leaned into the frame and sang hellos.

Together, they drove to the Leon Arms apartment complex, where Tony McDade had lived and had been shot, for the protest. There, during a moment of silence, Toyin whispered to Dani that she needed to get some fresh air and walked off. "Those were the last words I heard her speak," Dani said.

—

TOYIN HAD BEEN OUT OF TOUCH FOR ALMOST A DAY WHEN SHE POSTED ON Twitter that she had been assaulted. "I was molested in Tallahassee, Florida by a Black man this morning at 5:30 on Richview and Park Ave," she wrote. A man had given her a ride to get her things from New Life Methodist. After bringing her to his home, he had exposed himself to her while she showered. He offered to give her a massage, and she wrote that when he did, she froze. He was naked. When he fell asleep, she left and called the police, which TPD later confirmed. She tweeted a description of the man, his car, and his house. "I will not be silent," she wrote. "Literally wearing this man's clothes DNA all over me."

A friend, concerned by the thread, messaged her shortly after. Where was she? Toyin told her that she was at a library. The friend urged her to give a location so someone could collect her and bring her someplace safe. "All my friends are PTSD victims," Toyin replied. "I don't want to trigger anyone."

Dani and Ashley had been worried since Friday night. After the protest, Ashley and a friend had spent two hours looking for Toyin. And Saturday morning, at 8 a.m., Dani had driven to the police station to report her missing. Police told her she couldn't, as she was not an immediate family member. According to several activists in the city, small groups of people began looking for Toyin that day, talking to local businesses and combing through areas near where she was last seen. "We were out there from the jump," Dani said. Chancellor Crump, who had been a part of the early searches, said that the Twitter thread had terrified people. People took screenshots of it, posting it on their social media platforms or sending it to their group chats. "It spread like wildfire," he said.

After days of informal searches, on June 10 the police department released the official missing person flyer. As Toyin's name circulated on the internet that day, the TCAC coordinated its first official search party. These continued throughout the week, bringing over 50 people a night to scour the city. "It's a very close-knit community," Thomas said. "It doesn't take much for people to recognize somebody." This was especially true for the people who began to protest in the streets every day—many protesters became fast friends with people they had never met before in the city. "Even though we got to know each other in such a short amount of time, these protests have allowed people to become close. They're intimate," Guyton told me. "You get to know each other on a really personal level."

Thomas told me that the community of activists who showed up for Toyin had amazed her. "In a matter of two weeks we've met well over a hundred people," she said. TCAC created a GoFundMe to raise money that Toyin could use to find a new place and feel secure. "We created some posts saying that this money is for you to stabilize your life," said TCAC organizer Delilah Pierre. "Just in case she was looking at social media, she could see that we're not trying to put you back in that situation."

Between searches, they were still protesting. They were also organizing support for people who were arrested while protesting. "We're out here not only fighting on the front lines, but we also working behind the scenes to keep things going," Dani said. The police, meanwhile, had hardly interacted with the people leading the searches—even Dani, Ashley, and Tamra, the last people Toyin had been with, barely heard from the cops. Ashley said she spoke to one detective for about 30 minutes total over the course of the week. "Then I never heard from him again." Dani said she was in contact with a detective, too, but he eventually stopped responding to her. Indeed, the police viewed some of the community's decisions as more of a nuisance.

After the viral tweet that announced Toyin's disappearance, TCAC offered their number to people who felt uncomfortable speaking to the police, saying they would relay the information. This was, according to an official statement the Tallahassee Police Department published on June 16, "misinformation." TCAC had tried to forward them information they were gathering, including a text they received about a person claiming to be Toyin's kidnapper, but they say the police never followed up. "We all felt it was indicative of how little they cared," Pierre told me.

In a public statement posted to their website two days after Toyin's body was identified, TPD emphasized that though she had contacted them on June 6 about a sexual assault, they began searching for Toyin as soon as her "family reported her missing," on June 7.

They "intensified their efforts" to find her, reaching out to "known associates and victim advocate groups" and organizations that offer services to the homeless. "A missing person flier was widely disseminated, garnering national attention. A team of more than a dozen TPD investigators worked tirelessly to find Salau."

Community members who were a part of the search parties find this story difficult to believe, and they are still waiting for answers to what they say are urgent questions: Why did TPD wait three days to file a warrant to access Toyin's cell phone information? Why hadn't they checked the bus station security camera footage they would later use to arrest Aaron Glee, Toyin's killer? And why, as they canvassed neighborhoods and knocked on doors, did it seem as if the police hadn't talked to most of these people? "Once a week or so went gone and everybody was still searching for her, it really hit me . . . they're not trying," said Pierre. "Or if they are trying, they're really, really not trying hard enough."

Indeed, on June 12, Dani reached out over text to the detective she had spoken to that week. The detective asked what Toyin had been wearing at the protest when she walked off. "Was she wearing blue?" the detective asked, sending a photo. "Not at all. That picture was taken our second day protesting," Dani responded. "We've been at it for two weeks now." "OK," the detective responded. "Do you have a picture of what she was wearing?" Dani pointed out that Toyin had written in her Twitter thread that she was wearing her assaulter's clothing. By Sunday, the officer stopped responding.

Night after night, people gathered at Bethel AME, a church near FAMU, to search, only to return without any news. People began losing hope. "I'm no detective," Dani said, "but I know within every 24 hours after somebody has gone missing, the chances of finding them alive lessens a great deal."

—

ON SATURDAY, JUNE 13, A FAMILY FRIEND STOPPED BY TO CHECK ON VICTORIA Sims, known as Vicki, a 74-year-old white woman and longtime community volunteer, and noticed she had left her door ajar. The friend called the police. According to probable cause documents, family and friends told officers they had last heard from Sims two days before, though officers also noted that while that day's edition of the local paper was still on the front step, yesterday's was in her house. Sims's cell phone and car were gone. Family members told the police that sometimes Sims would give rides to a man named Aaron, who lived on Monday Road.

Officers traced Sims's phone and found it in her car, parked on Monday Road, less than 50 feet from a house that belonged to a man named Aaron Glee. After "breaching" the locked front door, according to the court documents, police found Sims's body. They noticed the room smelled like cigarettes, "as though someone was recently smoking." Officers sent out a search dog to see if Glee was hiding nearby. Instead, the dog found Toyin's body in the woods on an adjacent lot behind Glee's house, covered in leaves.

But on Sunday, the day after the bodies were found, Toyin's friends were still searching. They had felt there was reason to hope. Early that morning, the owner of Big Easy Snowball, an ice cream shop in a residential area in north Tallahassee, had contacted Octavia, saying that he was sure he had seen Toyin in his shop earlier that week. When Ashley and Dani drove over, he showed them camera footage. They were overjoyed. They were also eager to find out what she might be wearing, since they were unsure what clothes she had taken from the man, and they thought it was important for the people looking for her to know. Then Dani's eyes lit on her feet. "Those are my shoes," she said. The white Crocs.

She posted screenshots on Facebook. She wrote that the footage showed what Toyin was wearing, including the Crocs, and asked people to call the Tallahassee Police Department to relay the information. "I could barely press post when I got the call," she said.

For the community that had been rallying to give Toyin new hope, the news was shattering—but even more, the details of how she was found struck them as an injustice. "They found her by *accident*. They were looking for Victoria Sims, God rest her soul," Ashley said. "Do you know how disheartening it is for someone to find your friend by accident?"

—

HOURS AFTER TOYIN'S FUNERAL ON JUNE 27, IT RAINED IN TALLAHASSEE. A Florida rain, which pounded the rooftops and let out mutters of distant thunder over the dripping trees. For most of the city, it was a relief, clearing the humid air. For Toyin's friends, it was a minor delay in their plans. When the rain slowed, they drove out to Monday Road, where Toyin's body had been found two weeks prior.

Monday Road is a dark, narrow side street in a residential neighborhood, dense with shockingly green Florida brush. Like many streets in Florida, the trees, heavy with moss, form a canopy that blocks out the sky. In the dark it was hard to make anything out. But at the entrance to a long dirt road that ends in the home of Aaron Glee, who had confessed on June 20 to murdering Toyin, an array of candles made a warm dome of light that bounced off the signs and paintings placed in the trees and bushes.

"Have you seen all the young Black women coming up going missing?" Dani told me on the day of Toyin's memorial service. "There is something going [on] in this world, let me tell you, and they are trying to brush it underneath the rug like it is dust. And it's not.

This is not a game. This is not a joke. This is our lives being taken and this is real. They don't take us seriously."

Several people I spoke to told me that they're painfully aware of how little evidence police seem to need to fatally shoot someone on the street, or for a no-knock warrant, but how much evidence you need to get the police to address sexual assault. That they come out in force when people set off fireworks at a protest, but when a Black girl goes missing, they don't seem to be anywhere at all. The investigations into both Toyin's and Victoria Sims's deaths are still open. In a June 16 statement, the Tallahassee Police Department asked the public to "report any information they may have that could aid in Salau's original battery case or the double murder of Salau and Sims." But even in the aftermath, they say, the police have been careless—the photo they used to post their press release on social media is a screenshot of Toyin at the first protest she went to, on May 29. It catches her mid-speech, her face distorted, side by side with a posed picture of Vicki Sims. "She had so many pictures on the internet," Ashley said. "And they chose *this* one."

For now, Dani, Ashley, and the others who worked to get Toyin into a stable living situation are tired. They say they have a hard time sleeping at night. "I have to sleep with the lights on in my room," Ashley said. Dani moved out of the apartment where Toyin stayed with her. A friend who met Toyin on Twitter, also named Oluwatoyin, said she thinks about her all the time. "That could have been me. That could easily have been me," she told me. "I'm a young Black woman in America. When, as Black women, are we going to be safe?" Delilah Pierre said that it has taken time to work through the initial numbness she felt after Toyin's death was confirmed. She remembers watching, as though from a distance, as a group chat filled with people who had attended the search parties began to grieve. "A lot of people took it very personally," she told me. "Even though we didn't know Toyin very long, we saw her. We saw her potential."

On June 16, the day after police made her death public, over 100 people marched through the streets of Tallahassee, flanked by cars blasting music, to the spot on Monday Road where her body was found. They blanketed the road in votive candles and flowers and strung balloons to the trees. (At least six police cars followed them, their lights whirling.) The next day, hundreds more gathered at the steps of the capitol building and raised candles over their heads in her honor. (When some of the attendees set off fireworks, police approached, brandishing zip ties.) Activists across the country organized their own vigils for her: in Providence, Houston, New York City, Chicago, Washington, DC, Tampa, Boston, Miami, Atlanta.

At the same time, Toyin's story has already started to change Tallahassee, too. "The conversation has gotten a lot bigger as it relates to sexual assault. Even on the university level, there are greater conversations surrounding sexual assault that haven't happened before," Thomas told me. "How are we changing things on all levels? Not only the policy, but how are we changing both men and women's attitude as it surrounds rape?" For Dani, although Toyin is gone, she believes that there are lives that can be helped that are still here. She says that's what Toyin would have wanted. "She wanted to help people," she told me. "Somewhere out here in this world, you have a young man or young girl—and I say men because men are sexually assaulted, too—scared to come out and speak about what they've gone through, or who did it and why, and where, because of how much fear they instill."

Pastor Latricia Scriven says she has seen this herself. The day after she spoke at Toyin's vigil on the steps of the capitol, a young woman approached her and asked to speak with her in private. She told the pastor that she thought she might be in a predatory situation and that the vigil had given her the courage to reach out. "That is Toyin's legacy of life still at work," Scriven told me. "That is a part of her legacy."

ORIGINALLY PUBLISHED ON JEZEBEL, AUGUST 2020

"NO CHOICE BUT TO DO IT": WHY WOMEN GO TO PRISON

BY JUSTINE VAN DER LEUN

Tanisha Williams met Kevin Amos on December 29, 2002, the last day of his life.

Kevin, 19, lived with his parents, but sometimes visited his girlfriend and their infant daughter at the red-shingled apartment complex in Saginaw, a mid-Michigan city where Tanisha, 20, and her 32-year-old roommate, Patrick Martin, shared a basement two-bedroom. It was a dry, frigid winter, and a thin layer of snow coated the ground. That evening, Kevin came by for a drink. Patrick introduced him to Tanisha. She poured two glasses of Crown Royal on ice. Patrick had already been drinking all day.

In a bedroom, three of Patrick's children were watching TV. After a time, Tanisha headed to the kitchen. She heard Patrick's voice rise and stepped back into the living room, where she saw Kevin bleeding from the mouth and Patrick's hand raw, "the meat missing off [his] knuckle," she would later testify. The metal braces Kevin wore had torn Patrick's skin.

Kevin ran for the door, but Patrick ordered Tanisha to block his path. She obeyed. Patrick was six-foot-two and 215 pounds, always armed; he outweighed Tanisha, who is five-foot-seven, by nearly 100 pounds. Patrick dragged Kevin to the love seat, pistol-whipped him unconscious, stripped him naked, and kicked him repeatedly in the head and genitals. At some point—no one involved in the incident could remember the exact sequence of events—Patrick's cousin, Terrance Shepard, arrived. Months later, according to court transcripts, Patrick would confess to a girlfriend that he attacked Kevin because Kevin was "mean mugging" him—or looking at him the wrong way.

Patrick became concerned that Kevin would defecate. Trash bags and duct tape were fetched. Tanisha considered escaping, but she was barefoot and in a remote building with two male cousins. "I began screaming and yelling," she told me. Patrick grabbed her by her collar, picked her up off the ground, and slammed her against the wall, twisting the fabric of her shirt and pressing his fist into her neck. The drywall cracked, Tanisha recalled, and her throat closed. "Get down or lay down," Patrick said, his gun to her face. Then he released his grip.

Tanisha knelt on the floor. Following Patrick's instructions, she began to wrap tape around Kevin's head.

"How did you feel doing that?" Doug Baker, a prosecutor at the Michigan attorney general's office, asked her at trial years later.

"Terrible."

"Why did you do that?"

"Because I ain't have no choice but to do it."

—

TANISHA WAS BORN IN 1982 IN SAGINAW, ONCE THE CENTER OF A THRIVING automobile gear industry. A manufacturing decline in the late twentieth century led to urban decay. Crime rose; buildings sat vacant. Tanisha and her four siblings lived in a spartan 900-square-foot home. She remembers sitting in the house with only popcorn to eat, waiting for their mom to return. When Tanisha was a toddler, one of her mother's boyfriends dipped her feet in a tub of scalding water. Her mother grabbed her before the man could fully submerge her. A layer of skin peeled off her feet, which are still tough and scarred.

When Tanisha was about six years old, her mother was diagnosed with cancer. She married her boyfriend at the time so that he could act as guardian for her children while she underwent prolonged treatment. According to Tanisha, the man was addicted to drugs. Many days, Tanisha did odd jobs in the neighborhood and collected bottles to buy chips and hot dogs for her little brothers, walking alone in the dark to get food for them all. At night, her stepfather entered her room and offered her ice cream—especially appealing because she was often "super hungry"—in exchange for sex acts. Tanisha's mother later found out about the abuse. "You done slept with my husband," Tanisha recalls her mother saying. "I don't want to look at you." Tanisha was around 10 years old.

When Tanisha was 13, her family moved to Georgia. At 14, she became pregnant by a 26-year-old man. She had an abortion and was sent back to Saginaw to live with her father, whom she barely knew.

Sharon Sanders was married to Tanisha's father then; Tanisha and Sanders still consider themselves related. (In 2014, Tanisha's father was incarcerated for the sexual abuse of multiple young female relatives; he and Sanders later divorced.) Tanisha was "a hurt, angry, disturbed young girl," Sanders told me.

At 17, Tanisha had her first child. She told me that, at the time, she was "not able to be a mom." She left the baby with Sanders, dropped out of school, and bounced around, staying with adults who lived "the faster life." Until then, her only job had been at a fast-food restaurant; she remembers making less than $5 an hour. She found she could make $200 for oral sex: Her clientele, mostly drug dealers, paid big money as a show of status. Then, as now, Tanisha was slender, with close-cropped hair, high cheekbones, and an exuberant personality; she wore tight, sparkly dresses and organized parties where she provided sexual services. She found the work easy. "I was already introduced to being utilized by men."

At 19, Tanisha fell in love, got pregnant, and stopped doing sex work—but her boyfriend abused and cheated on her. She left her second daughter with him, slept in abandoned cars, showered in hotels, and sold small quantities of drugs to make money. She summarized that period in one word: "Survival."

Tanisha turned 20 in June 2002. Around that time, a step-cousin introduced her to Patrick Martin, a father of five who had recently separated from his wife. Patrick and Tanisha quickly decided to live together. Their relationship was pragmatic, not romantic: Tanisha needed a home, and Patrick wanted a domestic partner. They would split bills and maintain independent enterprises. "This is a perfect setup," Tanisha recalled thinking.

Tanisha soon learned, however, that Patrick had plans to act as her pimp, which she did not agree to. He started "jumping" her, once slapping her so hard that she saw "a flash of light." He beat and choked her, commandeered her finances, and demanded that she cook, clean, and care for his children. He confiscated her gun because, she would later tell a detective, "wasn't nobody else . . . gonna have a pistol but him."

"I could've went back to sleeping in cars, but my pride was too high," Tanisha told me. "I could deal with this while I stack some money somehow. I was figuring it out."

Then, just after Christmas, Kevin Amos stopped by for the glass of Crown Royal.

—

I FIRST HEARD FROM TANISHA IN FEBRUARY OF THIS YEAR WHEN AN ENVELOPE arrived in my P.O. box containing a response to a questionnaire that I had sent. "I'm at the beginning of my project," I had written in an accompanying letter to Tanisha and 548 others detained at Women's Huron Valley Correctional Facility in Michigan. "My goal is to write an article about why and how women end up in prison on murder charges. . . . I have attached 16 questions. . . . Tell me everything you think I should know about your story."

Tanisha later told me that she had lugged a communal typewriter to her bunk and fed the paper in. "I applied duct tape to the victim's head," she typed. "Only while a gun hung over me forcing me to assist. . . . In each step I thought I would die."

Over the last 21 months, I have sent my letter, questionnaire, and, as a modest show of appreciation, a crossword puzzle to thousands of women. New mail arrives weekly, and I'm sending out batches to more states. For the purposes of this article, I have analyzed, coded, and run statistics on 608 surveys: These are the responses from a total of 5,098 surveys sent to people serving time on murder and manslaughter charges in 45 state facilities for women in 22 states.

I started the project in late 2018, after I began investigating the story of Nicole Addimando, a young mother in upstate New York who had killed her abusive partner in what she said was an act of self-defense. In the course of that reporting, I came upon dozens of cases across the country in which a woman insisted that she had been trying to protect herself or a loved one, while the state countered that she was a cold-blooded killer and sought a harsh prison sentence. The stories were voluminous but anecdotal. Race and socioeconomic circumstances often played a prominent role, which is fitting given the history of gender-based criminalization in the United States. One of the first recorded cases occurred in 1855, when an enslaved Missouri 19-year-old named Celia killed her master, who had raped her since she was 14. She was represented by a slave owner, was convicted by a white male jury, gave birth to a stillborn baby, and was hanged.

I found limited studies, conducted in single prisons or states, consistently showing that up to 94 percent of people in some women's detention facilities experienced physical and sexual violence prior to incarceration. However, I couldn't find systemic data to support what experts told me, and what I witnessed while reporting: Women's prisons are populated not only by abuse and assault survivors, but by people who are incarcerated for their acts of survival.

About 230,000 women and girls are incarcerated, an increase of more than 700 percent since 1980. The female prison population is dwarfed by the larger population of more than two million imprisoned men, on whom conversations about mass incarceration center. For most people in prison, the criminal legal system has stripped away context and circumstance, leaving only a conviction on record.

Women must also navigate gendered binaries in a system designed by and for men: Offenders are violent, victims are docile; offenders kill, victims die. Female victims should fit a paradigm of innocence: a petite, heterosexual white woman with a clean record. Tanisha does not conform, though she has been unyieldingly victimized. But even women who do square with the paradigm struggle because they survived. "Lawyers say the only correct battered woman when talking about self-defense is a dead one," Sue Osthoff, cofounder of the National Clearinghouse for the Defense of Battered Women, told me. By engaging in violence in order to live, a woman cannot be a victim. Her survival itself becomes reason to condemn her.

In January 2019, I interviewed Rachel White-Domain, a post-conviction attorney for incarcerated survivors of domestic violence in Illinois. I pressed her, as I had others, for data on women in prison for defending themselves. White-Domain did not know the statistics. "If you're so interested, you could just ask the women directly," she said. That comment was the seed from which this work has grown.

In consultation with Thania Sanchez, then an assistant professor of political science at Yale University, I designed 16 questions to assess the abuse and trauma backgrounds and unique pathways of women into U.S. prisons. (Sanchez now works in the ACLU's data and analytics department.)

My two-page survey asked for demographic information—age, race, and sentence length—and posed qualitative questions. Those who speak of abuse are subject to doubt and skepticism; in a courtroom or a prison, they are often accused of trying to avoid consequences by making "the abuse excuse." To reduce any appearance of dishonesty, I did not ask "priming" questions about domestic abuse, sexual violence, or self-defense. Instead, my queries concerned the respondent's relationship to the person they were convicted of killing; the days leading up to the event; factors they believed contributed to their conviction.

I contacted the media liaison or public information officer at the department of corrections in the majority of states and requested a list of all women incarcerated on murder and manslaughter charges. Some officials denied my request, but many sent names: more than 1,000 in Florida, just 11 in Maine. Once I had the lists, I filled in each salutation by hand.

As in any data collection effort, the majority of recipients did not respond. People in prison face particular hurdles in corresponding: censorial mailroom staff, fear of retribution from staff and administration, distrust of media, legal concerns, and constraints that include count times, chow calls, and lockdowns. Incarcerated respondents had to pay for a 55-cent stamp despite working jobs that were unpaid or, in some states, began at less than 4 cents an hour.

Mental illness, which affects nearly half of those in prison, was another hurdle: "Every prosecutor describes women convicted of murder as cunning, diabolical, monster, and evil," Kwaneta Harris wrote from Texas. "I've yet to encounter these 'monsters.' Although I've met plenty of women with mental illness, untreated and undiagnosed . . . the ones who you really need to talk to are too mentally damaged to talk to you."

Still, 604 cisgender women, one transgender woman, and three transgender men replied. They said that they wanted their stories told: sometimes anonymously, sometimes with their identities central. Many wanted to make known their own experiences and those of others. One woman serving life in prison, who had transferred between state and federal facilities for 17 years, repeated a common sentiment: "I have lived amongst thousands of different women prisoners from different countries, ethnicities, and cultures. I have probably come across six of these women whom I would even think were murderers. Many of us were defending ourselves, with the wrong people at the wrong time, taking cases for someone else, or not guilty completely."

The racial breakdown of responses roughly mirrored that of the larger female prison population, which is majority white but disproportionately women of color. My respondents were 53.7 percent white; 32.7 percent Black; 8.8 percent Hispanic; 3 percent American Indian and Alaska Native; 1 percent mixed race; and 0.8 percent Asian. Their ages spanned 18 to 83. The median age was 43, which skews older than the median age for all incarcerated people, 36. Nearly 30 percent were serving life sentences, including life without the possibility of parole. Overall, the average sentence, including life sentences (weighted at 100 years), was 55 years.

Seventy-two percent of respondents had been represented by state-appointed defenders, meaning they most likely qualified as indigent. A 2015 Prison Policy Initiative analysis found that the median annual income for a woman prior to her incarceration was $13,890, 58 percent of the income of a woman who was not incarcerated and 34 percent of the income of a man who was not incarcerated. "No money for attorneys, unfamiliar with the legal system," one respondent summarized. Another woman wrote that her case was her lawyer's first-ever murder trial. One wrote that the judge fell asleep and her attorneys "said it was okay."

Sixty percent of respondents reported experiencing abuse—physical, sexual, emotional, or some combination of the three—before entering prison, much of it during childhood, while only 9 percent reported no abuse. (Thirty-one percent did not give enough information for a determination.) The fact that a majority of respondents had been abused suggests a nexus between abuse and incarceration for women. Abuse is also most likely underreported.

"I have yet to meet a person who hasn't been sexually or physically abused," wrote Kwaneta from Texas. Later, Kwaneta told me that she had been kidnapped and gang-raped as a child, and the perpetrators were never charged. (Her mother confirmed this.) Kwaneta is incarcerated for killing a boyfriend who, she told me, abused her. Her attorney, she said, didn't want her to "disparage" the dead at trial. She did not push back. "The stigma and shame of allowing myself to continually accept abusive behavior is stronger than the shame of being a convicted murderer."

Having a female body had opened the respondents up to harm long before they were considered to have harmed others. They reported being raped at gunpoint, raped while being driven home from babysitting, raped by fathers, stepfathers, brothers, grandfathers, cousins, uncles, foster relatives, and sometimes sisters and mothers. "Raised for sex," wrote one woman. "Ruined before I had a chance," wrote another. "Not sure how I got off the beaten path, but I was molested by a sheriff from age 7 to about 9 years old, think that pretty much did it."

They wrote, too, of enduring physical violence: forced to kneel "on uncooked grits until my knees were bloody," kicked "with steel-toed boots," hit in the head with "a tire wrench." They wrote of crushing poverty and upheaval: One woman had used "the bathroom outside for years, shower[ing] in the water hose." Another said that when she was 19, she lived with her drug-addicted mother in a "condemned house" in the days leading up to the killing she committed when she fought off an attempted rape.

A number of life factors usually converged—many beyond a person's control. I received a letter in purple crayon from a 32-year-old Black woman who wrote that "the judge . . . gave me to [*sic*] much time and I was 11 when I did the murder." I looked up her story, which was public record: As a child, she had been raped, abandoned, neglected, cycled through abusive foster homes, and diagnosed as mentally ill and developmentally disabled. Before her twelfth birthday, she stabbed a stranger in the heart with a kitchen knife. For nearly two years, she lived in an isolated cell in an adult jail before being sent to a facility for children with mental illnesses. When she turned 18, a judge sentenced her to 18 to 40 years.

Childhood abuse and neglect rippled out into adulthood. More than 40 years of research and multiple studies show that abuse begets abuse, and that sexual victimization in childhood raises the risk of sexual victimization in adulthood. Children who have been objectified and betrayed can have issues with trust, and find it difficult to navigate adult relationships. My survey found that people abused as children were more than twice as likely to report abuse as adults: Nearly 75 percent reported revictimization, versus 33 percent among those who didn't report abuse as children—meaning that those with abuse histories are more than twice as likely to be victimized in their later years.

Forty-three percent reported experiencing intimate partner violence, nearly double the rate of the general populace. Of those, 41 percent—nearly 18 percent of all respondents—said they were in prison for killing a romantic partner. They wrote that they had killed men who had broken their ribs, backs, knees, skulls—men on whom 911 had been called (in one instance 300+ times). Men who had knocked them awake and said, "I feel like beating a bitch up." According to the Centers for Disease Control and Prevention, roughly half of female homicide victims in the United States are killed by a current or former male partner.

"That morning he said, 'One of us is going to die today,'" wrote a woman who had been brutalized for decades. "I just snapped."

Respondents found it hard to prove that they had been abused, because domestic violence and rape are private violations, usually without witnesses. But even if they had evidence, they faced another problem: There was a dead body, and it wasn't theirs.

Jema Donahue wrote from Missouri, asking me to contact her mother, who shared legal files with me. The files showed that in 1999, Jema, then 13, was raped by a 20-year-old man. Jema's family was devastated by their attempts at seeking justice. The man's mother was the town mayor, and police and prosecutors, Jema said, resisted investigating and prosecuting. The perpetrator eventually pleaded guilty to statutory rape, though Jema said he had used force, and she had experienced bleeding and physical pain for days. The rapist's friends harassed the family. Jema was bullied and switched schools. Later, Jema fell into abusive relationships. Her husband physically, sexually, and psychologically terrorized her for eight years—attacks that she sometimes reported to authorities. In 2017, Jema's mother took out an order of protection. Within two weeks, Jema's husband broke into the family home and attacked Jema. Jema shot him. A judge sentenced her to 10 years, noting that "everybody stayed in the relationship" and "somebody died."

"It was supposed to be me," Jema wrote to me. "I never thought I would be the survivor and he 'the victim.'"

—

OF MY RESPONDENTS, 30 PERCENT REPORTED THAT THEY, LIKE JEMA, WERE IN prison for trying to protect themselves or loved ones from physical or sexual violence. If my findings are representative of the population incarcerated for murder and manslaughter in all U.S. women's detention facilities, they suggest that, conservatively, more than 4,400 women and girls are serving lengthy sentences for acts of survival, and that there are most likely others in similar circumstances serving time on lesser charges. While many claimed straightforward self-defense against a romantic partner, others wrote of trying to survive in ways that exist outside the typical ideas of gender-based violence.

"In a lot of cases, the women are not blameless," Carol Jacobsen, director of the Michigan Women's Justice & Clemency Project, told me. "They played some part, many times under terrible duress. Many times, they didn't have a whole lot of choices." A woman was often convicted for a man's actions or for her involvement with a man who had committed a crime. Thirty-three percent of respondents said that they had been convicted of committing their crime with a male partner, and 13 percent said that they had been convicted of committing their crime with their abuser—often, like Tanisha, under duress.

"I should get prison time, but not 50 years," wrote a woman who had driven a car for her abusive boyfriend after he'd killed a man during a burglary. "Fifty years for driving a car under duress . . . with the threat always there to kill me and take away my kids for talking or leaving him."

Motherhood figured prominently in a woman's decisions; 64 percent of my respondents said they were parents, 22 percent said they were not, and 14 percent did not answer the question. "One topic . . . affects the majority of female prisoners: Our children," wrote Kwaneta in Texas. "We care about our children. Bars and razor-wire don't erase motherhood."

Some women said they were in prison for trying to protect a child. A white woman was serving 40 years in the South for shooting the man she said had beaten her daughter and raped her toddler grandson, and a Black woman was serving life in a mid-Atlantic state for shooting a man she witnessed molesting her five-year-old son.

Other women were convicted because they had not protected a child. In Michigan in 2010, Corrine Baker, a 25-year-old survivor of childhood sexual abuse, threw her body in front of her four-year-old son in an attempt to bear the brunt of her boyfriend's attack. Corrine and her son were hospitalized after the beating. The boy died, and Corrine gave an interview to a local TV station; in the video, she has two black eyes and wounds dotting her face. She is serving 13 to 30 years for her failure to save her child.

In Florida, Mary Rice wrote that she was forced to accompany a man on a multistate killing spree, during which she was starved, drugged, and raped. She never tried to escape, she told me, because her kidnapper knew where her mother and three children lived. In the end, he shot himself in the head, and police took Mary, who had injuries across her face and body, into custody. Mary was prosecuted. The assistant state attorney told the jury, "She wanted a bad boy and she got one." Mary was sentenced to life in prison plus 30 years. "I just lived through it," Mary wrote to me. "The state needed to blame someone. I was the person they blamed."

Mary went to trial, but about half of respondents pleaded guilty—often because they were marginalized or did not trust their lawyer. I received a response from an Indigenous woman in the Northwest, who said she had pleaded guilty to murder, despite acting in self-defense against a man who was "abusive physically, emotionally, verbally, and mentally," had raped her daily, and threatened to keep her child if she left. I called her public defender. He told me the woman had a weak case and added that in court, women often played the victim. "Women get a lot of fuckin' benefit of the doubt," he said.

"I see a lot of pleas," White-Domain, the Illinois attorney for incarcerated domestic violence survivors, told me. "People who have lived their whole lives not being believed—if they are women, people of color, maybe also a sex worker, maybe a drug user—think they won't be believed by a jury. And they are correct." In my data set, those who went to trial received a sentence nearly two times higher than those who did not, and were approximately five times more likely to receive a life sentence. Black women, whether they went to trial or took a plea, received a sentence that was roughly 10 percent longer than all others.

Tanisha, for her part, pleaded guilty in 2010 to second-degree murder while held in Saginaw County Jail, a situation she found herself in largely because she had tried to make amends for Kevin's death. "I was ignorant to the system and its workings and it worked against me on every facet," she told me. "The rules weren't articulated for a person with zero understanding to understand."

—

IN LATE DECEMBER 2002, PATRICK WRAPPED KEVIN IN A BLANKET AND PUT him in a closet. Within days, aided by two associates, he removed the body and dropped it down an embankment on the outskirts of town, above a river. About three months later, on March 23, 2003, during the spring thaw, two fishermen came upon Kevin's corpse. On March 24, Kevin's family was informed. Police made no arrests. The case went cold.

Two weeks after the killing, Tanisha and Patrick moved to a more isolated apartment. Tanisha had no car and no money. She slept during the day, when Patrick was gone, and kept a gutting knife tucked in her waistband. "Something was gonna happen bad for someone," she told me. "Most likely me." Within two months of Kevin's death, Tanisha gathered some $200 and four outfits and fled to a roadside motel.

"I sat on that ugly bed," Tanisha said. "I had to make a decision." She was 20, without a home or anyone she trusted enough to talk to. She had been involved in a murder with a known drug dealer who threatened to kill her and knew where her family lived. Tanisha decided her only chance was to stay moving, and be quiet. "I pushed what happen to me to the deepest depths of consciousness," she recalled. For the next two years, she fell into a cycle of sex work, drug use, and dealing. She lived on the run, involved with men who hurt her.

"I didn't dream," she said. "It was just darkness."

But in 2005, Tanisha was pregnant and determined to get clean and make a change. In November, she gave birth to her third daughter, whom she named Hon'Esty. As Tanisha raised Hon'Esty, she began to see a man around town who bore an unnerving resemblance to Kevin. She thought constantly of Kevin's mom. She feared intensely for her daughter's safety, concerned that by having a fixed address, they were a target. Each time Tanisha entered her home at night, she left her child locked in the car, on speakerphone with her sister, as she swept the place with her gun cocked. Only once it was clear would she carry Hon'Esty inside. Later, two people testified at trial that Patrick said he was thinking about killing Tanisha; he worried that she would talk to police and had unsuccessfully attempted to convince her to meet up with him several times.

Tanisha made a New Year's resolution for 2009: She would tell the truth. That February, she called an attorney she'd heard of named Steven Snyder. Tanisha told Snyder that she wanted to talk to police but needed immunity. Snyder called Lisa Speary, a Michigan State Police detective who had been handed the cold case in 2006. On February 17, 2009, Snyder, Tanisha, and a Saginaw County prosecutor signed a proffer, a written agreement that allows a person to speak of a crime with the assurance that their words won't be used against them in criminal proceedings. After an initial proffer, a client typically provides a piece of evidence that allows authorities to determine their value; then a better deal is negotiated. But Tanisha signed a single agreement that offered her no protection from charges. Then she gave Speary a two-hour interview, the first time she had ever spoken of the crime.

I showed the proffer letter to David Moran, cofounder of the Michigan Innocence Clinic. He characterized it as "lousy." I asked Moran if it was unusual for an attorney to allow a client to give a full statement to police without discussing the content beforehand. "The lawyer would want to hear the client's story before having her tell it in a proffer," he said. (Snyder, who no longer practices law, did not respond to my interview request.)

Over the next months, Tanisha worked with Speary. They met in person at least five times, Tanisha underwent two polygraph exams, and she led Speary through the crime scene. She told Speary that Patrick "had this rage . . . he frightened me . . . he was more dangerous than the average guy that I have been involved in . . . he was dangerous without maybe knowin' it." And she told Speary that following the killing, she had "spiraled like super down." Tanisha explained that she was coming forward to give closure to Kevin's family—and to get closure for herself. "I just wanna apologize to the family . . . for even takin' this long to get the strength to tell 'em what happened."

Though Tanisha remained terrified of Patrick, she was also liberated. "I started having joy in my life," she told me. She thought Snyder and Speary would protect her, and that a prosecutor would recognize her courage and commitment to justice.

"Why did you think that?" I asked.

"Because this is America. . . . I thought if I was just honest, the truth was going to set me free based off American values."

Speary interviewed witnesses widely and collected evidence. Then, on June 22, 2009, Patrick called 911 to report that his girlfriend, a 44-year-old nurse named Debra Kukla, was unconscious in the garage. When police arrived, they found Kukla bludgeoned to death.

Tanisha told me that Speary had called her days before Kukla's death; on the call, Speary asked for Tanisha's permission to tell Patrick that she had been speaking with police. Petrified, Tanisha refused. She theorized that police had nonetheless confronted Patrick, and that "he thought [Kukla] was the one who talked." Tanisha turned 27 the day Kukla died. She recalled that Speary called her that morning "in a panic," saying, "he did it again," and advising her to find a safe place, because police didn't know where Patrick was. (According to Tanisha, no one working for the state offered to provide her with security.) "I feel if I had never entered the agreement, Deb would be alive," Tanisha told me. "Every year at my birthday I think about her, and I think that I survived." Speary, now retired, didn't respond to requests for an interview.

Tanisha also believed that her incarceration was related to the fact that Kukla was white. "They were like, 'We have to get him off the streets by any means,'" she told me. "I believe they conjured up they moves, and I was the casualty."

The attorney general's office could not provide me with the exact date it took the case. But the office's press secretary said it was approximately September 2009—nearly seven years after Kevin's killing, but only months after Kukla's. On September 11, 2009, Patrick was arrested by an off-duty police officer who said he had witnessed him robbing a 7-Eleven. In March 2010, Tanisha was arrested at her job at a golf course. She was charged with first-degree murder and booked into Saginaw County Jail. Hon'Esty, then four years old, went to stay with relatives. "She just was always looking for me [after that]," Tanisha told me. "She would sit at her desk at school writing me letters."

The court appointed William White, a private attorney with a county contract, as Tanisha's defender. According to a letter that White later wrote to the judge, he was constrained by a "$1,000 cap" on his legal work for Tanisha. He billed for 36.5 hours, which means that, unless he was granted a fee extension, he was paid $27.40 per hour—a rate that diminished the more he worked. A homicide case, according to Moran, of the Michigan Innocence Clinic, "is the legal equivalent of performing brain surgery. It's complex and requires a great deal of skill to be able to do it right."

A 2008 National Legal Aid & Defender Association report on Michigan's indigent defense systems studied sample counties and found that none of their public defender services were constitutionally adequate. The fixed-rate system, which also exists across the United States, created "a conflict of interests between a lawyer's ethical duty to competently defend each and every client and her financial self-interests that require her to invest the least amount of time possible in each case to maximize profit," according to the report.

Since 2011, legislative efforts have led to indigent defense reform in Michigan, and in 2019 Saginaw opened its first public defender's office. Steve Fenner, a former prosecutor who heads the new office, told me that $1,000 to work a homicide was "insane," and that the previous system meant that attorneys "basically lost money" on major cases. But Fenner didn't understand why Tanisha needed an attorney at all. "Why was she charged though? I don't get it. That part is real baffling. She cooperated to solve a cold case, then the attorney general's office turns on her? I'm virtually speechless."

Tanisha had little interaction with White, her attorney. One day, White brought Sanders, Tanisha's former stepmother, to a courthouse meeting. Sanders told me that she had arrived for what she believed was a hearing and was surprised to be taken aside by White, who asked that she convince Tanisha to testify at Patrick and Terrance's trial.

Deputies brought Tanisha, in an orange jumpsuit, into the room. Sanders started to cry. "I say, 'This has been going on for a long time, and it need to come to an end . . . release all of us from all of this . . . whatever it is that they seeking, you need to give it to them.'" Sanders had no experience with the legal system and was caring for one of Tanisha's daughters. She told me that she believed she was helping Tanisha and knew nothing about a plea deal. "I thought they was gonna let her come home . . . because she had gave them what they wanted."

Tanisha, after two hours of Sanders's exhortations and nine months in jail, agreed to testify in exchange for a second-degree murder plea that she originally believed was 20 years flat, not 20 to 40 years. "I took the plea 'cause I was sick of being in there, hearing my momma beg me," she told me. White did not respond to a request for comment.

In January 2011, Tanisha testified for the prosecution. For the attorney general's purposes, Tanisha needed to inhabit contradictory roles: moral and credible enough for a jury to trust, but blameworthy and sufficiently deplorable to exist as an extension of the man who killed Kevin and to therefore merit her own conviction. Doug Baker, the prosecutor, characterized Tanisha and the others as "jackals."

Baker questioned Tanisha as a key witness over two days, using her testimony as a basis for a larger narrative. Then in his closing arguments, Baker alternately diminished and commended Tanisha. He told the jury that when Tanisha met Patrick, at 20, she was living "a wasted life . . . she is prostituting herself. She has children that she's not living with." Tanisha was "not a very reflective or thoughtful person." But she "had some conscience" and had come forward. Tanisha had acted under duress, he said, but "that's not a defense to homicide. . . . The law says, no, you can't do that. You've got to resist. You've got to—you've got to take your chances."

In Michigan and many other states, the reason duress cannot be used as a defense for homicide is based on British common law, as summarized by Matthew Hale, a Puritan jurist, who wrote in an influential treatise in the 1600s that even "if a man be desperately assaulted, and in peril of death . . . he ought rather to die himself, than kill an innocent." Hale also put forth other enduring writings and decisions. In one, he sentenced women to death for sorcery, one precedent used to justify the Salem witch trials. In another, Hale stated that by signing a marriage contract, "a wife hath given herself in this kind unto her husband, which she cannot retract." A marital exception to rape law, based largely on Hale's work, existed in England and Wales until 1991 and in North Carolina until 1993. The criminal legal system still treats spousal rape with leniency.

"[Tanisha's] choice was she should die," Jacobsen, of the Michigan Women's Justice & Clemency Project, told me. "Or prison. That's it. It doesn't matter that he's going to kill you. . . . You let that happen."

"Black women can be disposed as an object of punishment in order to demonstrate that the system works," Alisa Bierria, an assistant professor of African American studies at the University of California, Riverside, and a cofounder of Survived & Punished, an organization that supports incarcerated survivors of gender-based violence, told me. "This is what the system does, that's what it is: It has to perform justice in order to have good copy, but it performs that on the backs of Black women all the time because nobody is interested in the full dimensionality of their story as a human being."

"Surviving all of that shit, I can't even believe this is my life for real," Tanisha told me. "I really be like, 'God, what's up? What is all of this about?' I love taking care of the earth, I love kids . . . I got so much suffering and I never did nothing." The authorities, she said, "didn't and don't care what happened to me. No one don't even know how I lived. . . . They got me in here and really don't know what that man did to me that night."

—

IN OCTOBER, PATRICK MARTIN CALLED ME FROM KINROSS CORRECTIONAL FAcility in northern Michigan, where he is serving life for armed robbery and two murders. (He pleaded no contest to killing Debra Kukla after a jury convicted him of killing Kevin Amos.) Patrick said that he had bipolar disorder with psychotic features, was "a full-blown alcoholic," and had been institutionalized in 2007 and 2008. He described years of "madness," and told me that his 2009 arrest was "almost a relief." Patrick could not express why his mental illness and alcoholism manifested in lethal violence, including toward women, but he could speak clearly to why Tanisha and his cousin Terrance had participated in Kevin Amos's murder. "I made them do that," he said.

Would he have killed Tanisha, I asked, if she hadn't obeyed?

"I always had them guns and stuff. . . . My own mom would be scared. At that time, [Tanisha] was terrified. . . . I told them, 'If you don't do it, I'ma do you.' So they did it, but it was out of fear. They wasn't with the plan."

Eight days later, I called Doug Baker, the prosecutor in Tanisha's case, now chief of criminal enforcement and quality of life for the city of Detroit. We spoke at length about his reasons for charging and sentencing Tanisha to decades in prison, considering the role she played in the actual crime and in helping authorities.

Baker explained that when a witness has accepted a lengthy sentence, it can benefit the prosecutor who puts her on the stand. "Part of what goes into making a witness like [Tanisha] credible is that she's paying for what she did," Baker said. "And the jury hears that. If a jury takes somebody that gets . . . probation or whatever, that's argued to the hilt by the other side that they would say and do anything to get this sentence." This suggests that a prosecutor, who has nearly total discretion in charging decisions, may be incentivized to seek the most serious charges that ultimately carry extensive sentences.

Baker also argued that Tanisha's sentence was a form of justice for Kevin and his family. Tanisha, he said, was "the one that actually in that sense literally took the life. She made sure that he wasn't going to get those last breaths."

At the 2011 trial, the medical examiner testified that Kevin died of asphyxiation, because there was duct tape on his mouth and liquid in his lungs. He also testified that two quarters were found in Kevin's stomach and a bag of powder was found in his mouth, and that a dead person cannot swallow—so Kevin was alive when the quarters were put into his mouth.

Since her first interview with Speary in 2009, Tanisha has denied any knowledge of the bag or coins. At trial, a confidential informant for the state, a drug dealer, testified that he briefly stopped by the apartment on the night of the attack and saw Kevin bound on the floor. Within days, he helped dispose of Kevin's body. In August 2003, eight months after the murder, the informant was in jail, looking to cut a deal to get out two months early. He told authorities about Kevin's killing, and in order to prove that he was reliable, offered a detail that "nobody else would know except for somebody that had firsthand information," he testified. The detail was that Patrick "said that he put 50 cents in [Kevin's] mouth to make it look like a drug transaction."

When I spoke with Patrick, he didn't recall anything about the coins. Still, it seemed possible that Patrick may have removed and replaced the tape over Kevin's mouth, the final act that killed him. (Terrance Shepard, serving a life sentence in a southern Michigan prison, declined to comment for this story.)

At trial, Baker told the jury: "It's 20 to 40 years out of her life. She—at least 20. It could be more." On the phone, after we had discussed Tanisha's story, Baker said, "I think that she would be a good subject for being paroled."

At Tanisha's March 2011 sentencing, Baker provided her with a letter on attorney general's office letterhead. He wrote that she had participated in the homicide "after being threatened."

"I can attest to Ms. Williams' complete cooperation, candid testimony and remorse for her role in the crime," Baker wrote. "There is no question that Ms. Williams, as a Prosecution witness, was of invaluable assistance in bringing the other Defendants to justice. . . . We believe that Ms. Williams can be rehabilitated and someday live a law-abiding life."

However, just before her arrest in 2010, Tanisha was stable and productive. She was sober, raising her daughter, and working three conventional jobs. "I was a proud, tax-paying citizen," she told me. "I was on a positive path."

I read Baker his 2011 letter. I asked him what Tanisha's lengthy sentence was supposed to accomplish.

"It has a deterrent value, as well as a punishment value, as well as a rehabilitative value," he said.

But neither Tanisha's nor Patrick's incarceration seemed to produce any constructive change in their closest circles. The cycle of gender-based violence and mental illness has continued. Patrick's son, Patrick Allen Martin Jr., was 11 when Kevin was killed. As a young man, he was sent to prison. "He was different when he came home," his sister would later testify. He lived in a facility for people with mental health issues.

In 2019, Patrick Jr., then 27, shot MoeNeisha Simmons-Ross, Tanisha's 26-year-old niece and a mother of three who was also pregnant with Patrick Jr.'s child. MoeNeisha and the baby died. Her brother told local media that MoeNeisha's other children "were in the apartment and they saw what happened."

"The system is a freaking violence-producing factory," Bierria, the Survived & Punished cofounder, told me. "Relentless."

—

A THEME EMERGED IN MY RESEARCH: AS IN TANISHA'S CASE, INCARCERATED women, before their involvement in the legal system, were regularly disregarded or damaged by state systems, from Child Protective Services to schools to police. Families and individuals in desperate situations did not have access to quality services, or to sustained services—and were often scared of seeking outside help. "In the Black community you don't go to the cops," Tanisha told me. "You just solve your own shit. And then with a crime, the code is you never talk, don't ever talk, and I see why."

Sandra Brown, a Black woman in Illinois, wrote of years of terror, beginning in childhood. She was beaten, spit on, and bullied at school. When she fought back, she was expelled, labeled "aggressive" and "dangerous." At home, she was whipped. After she explained why she had welts on her legs to a trusted teacher, she says she "paid terribly for it." Sandra was later a victim of domestic violence and rape. As a young mother, she was arrested for killing a woman in what she says was an act of self-defense. She was sentenced to 22 years in prison. "The tragedies we suffered as little girls and young women in a sense 'groomed and doomed' us to this current state of modernized slavery," Sandra wrote me. "I have been the recipient of acts of violence since I was a child, and the law was virtually nowhere to be found. But the one time I fight back because I am afraid for my life, I am now a 'violent' offender."

Bierria observed that stories of gender-based criminalization were the result of the legal system's design and function. "It's not, 'Oh what a sad story, the prosecutor didn't care, the defense was bad,'" she said. "Those things animate the system we have. . . . What you see are formalized acts of profound, life-ending violence."

What was the alternative, particularly for women?

"We have to build it," Bierria said. Across the country, she said, people have long been engaging in concentrated, community-based anti-violence and transformative justice work, training, and education. "There's no magic answer that will get us where we need to go. All we have is us. . . . I think people are on it, and therefore I think there is a chance."

The respondents to my survey offered various solutions, including childhood intervention, decriminalizing poverty, treating mental illness and addiction, effective protection for sexual and domestic violence victims, changing incentives for police and prosecutors, engaging offenders and victims in restorative justice processes, and capping sentence lengths. One woman suggested that people be permitted to tour prisons and jails: "Allow the public to see who is in their prisons."

Tanisha, for her part, wanted to be seen—as a way to advocate for herself and others. I asked her why she responded to my letter in the first place. "I know everything I've been through," she said. "This matters for women. . . . I felt that every little bit helps."

Tanisha understood that at each turn she had been failed: as an abused child, a homeless teenager, a traumatized young mother, and perhaps most significantly by a state apparatus that reduced and exploited her story and good-faith efforts to bring closure to Kevin and his family. But she remained hopeful. She never complained about the abysmal conditions at Huron Valley, with its endemic black mold, repeated scabies outbreaks, and waves of Covid. She meditates, does yoga in the mornings, pores over legal papers, and works disinfecting the facility at night.

I asked Tanisha how she remained so persistently optimistic. "It's some favor," she said. "Something in my spirit that sustains me." She thought of her efforts to shine a light on injustice in biblical terms: as a mustard seed. In Scripture, the mustard seed is the smallest of seeds, but, when sown, becomes a tree, its branches filled with birds. "As long as we got a mustard seed, we can make it grow."

ORIGINALLY PUBLISHED IN THE *NEW REPUBLIC*,
IN PARTNERSHIP WITH THE APPEAL, DECEMBER 2020

THE GOLDEN AGE OF WHITE-COLLAR CRIME

BY MICHAEL HOBBES

1. A SLOW-MOTION LOOTING

Between 2018 and 2020, nearly every institution of American life has taken on the unmistakable stench of moral rot. Corporate behemoths like Boeing and Wells Fargo have traded blue-chip credibility for white-collar callousness. Elite universities are selling admission spots to the highest Hollywood bidder. Silicon Valley unicorns have revealed themselves as long cons (Theranos), venture-capital cremation devices (Uber, WeWork) or straightforward comic book supervillains (Facebook). Every week unearths a cabinet-level political scandal that would have defined any other presidency. From the blackouts in California to the bloated bonuses on Wall Street to the entire biography of Jeffrey Epstein, it is impossible to look around the country and not get the feeling that elites are slowly looting it.

And why wouldn't they? The criminal justice system has given up all pretense that the crimes of the wealthy are worth taking seriously. In January 2019, white-collar prosecutions fell to their lowest level since researchers started tracking them in 1998. Even within the dwindling number of prosecutions, most are cases against low-level con artists and small-fry financial schemes. Since 2015, criminal penalties levied by the Justice Department have fallen from $3.6 billion to roughly $110 million. Illicit profits seized by the Securities and Exchange Commission have reportedly dropped by more than half. In 2018, a year when nearly 19,000 people were sentenced in federal court for drug crimes alone, prosecutors convicted just 37 corporate criminals who worked at firms with more than 50 employees.

With few exceptions, the only rich people America prosecutes anymore are those who victimize their fellow elites. Pharma frat boy Martin Shkreli, to pick just one example, wasn't prosecuted for hiking the price of a drug used to treat HIV from $13.50 to $750 per pill. He went to prison for scamming investors in a hedge fund scheme years before. Meanwhile, in 2016, the CEO whose company experienced the deadliest mining disaster since 1970 served less than one year in prison and paid a fine of 1.4 percent of his salary and stock bonuses from the previous year. Why? Because overseeing a company that ignores warnings and causes the deaths of workers, even 29 of them, *is a misdemeanor.*

Construction magnate Bruce Karatz provides an infuriating case study of how the criminal justice system treats wealthy defendants. In 2010, Karatz was convicted of failing to disclose in a financial statement that he had secretly "backdated" his stock options (think Biff with the Sports Almanac in *Back to the Future II*) to boost his pay by more than $6 million. Prior to his sentencing hearing, his lawyer submitted letters of support from former mayor of Los Angeles Richard Riordan and billionaire philanthropist Eli Broad. Prosecutors recommended six and a half years in prison; the judge gave Karatz five years' probation and eight months of house arrest in his Bel Air mansion. After two years, the judge terminated the remainder of the sentence. Karatz later received a civic award from the *Malibu Times* for volunteer work he did to make a good impression for his sentencing hearing.

Country-club nepotism and Gilded Age avarice are nothing new in America, of course. But the rich are enjoying a golden age of impunity unprecedented in modern history. "American elites have become more brazen than they were even five years ago," said Matthew Robinson, a professor at Appalachian State University and the author of several books on "elite deviance"—all the legal and illegal social harms caused by the wealthy.

Elite deviance has become the dark matter of American life, the invisible force around which the country's most powerful legal and political systems have set their orbit. Four members of the Sackler family, the owners of Oxycontin maker Purdue Pharma, have retained the services of former SEC head Mary Jo White as their personal lawyer. Epstein's dinner party guest lists included Harvard professors, billionaire philanthropists and members of political dynasties in at least two countries. In 2017, the pharmaceutical company Novartis spent about 14 percent of its annual lobbying budget on payments to a shell company controlled by ex-Trump lawyer Michael Cohen.

And this clubbiness has human costs. Tax evasion, to pick just one crime concentrated among the wealthy, already siphons up to 10,000 times more money out of the U.S. economy every year than bank robberies. In 2017, researchers estimated that fraud by America's largest corporations cost Americans up to $360 billion annually between 1996 and 2004. That's roughly two decades' worth of street crime every single year. As the links between corporations and regulators become increasingly incestuous, the future will bring more crude-soaked coastlines, price-gouging corporate behemoths and Madoff-style Ponzi schemes. More hurdles to suing companies for poisoning their customers or letting bosses harass their employees. And more uniquely American catastrophes like the opioid crisis and the price of insulin.

Perhaps the greatest myth about white-collar crime is that Americans struggle to understand it—as if chemical companies toxifying rivers or insurance executives gouging their customers fail to stimulate our moral intuitions. In fact, surveys consistently show that the vast majority of the population considers white-collar crime more harmful than street crime and powerful offenders more odious than common criminals.

Those intuitions are correct: An entrenched, unfettered class of superpredators is wreaking havoc on American society. And in the process, they've broken the only systems capable of stopping them.

2. AN INCREASINGLY DESPERATE PANTOMIME OF LEGAL ENFORCEMENT

Every year, at branded cocktail receptions and bloated buffet breakfasts, government agents spend two days hobnobbing with the tax-haven attorneys they spend the rest of the year investigating.

The OffshoreAlert conference takes place in Miami each spring, in London each fall and in "key offshore jurisdictions" all year round. Officially, participants come to discuss "wealth creation, preservation and recovery." Less officially, the tax lawyers come to learn what the feds will crack down on next year. The government investigators come to fish for future jobs. Imagine a yearly picnic where sheriffs give drug dealers tips on hiding baggies from pat-downs and leave with a new set of endorsements on LinkedIn.

In person, the conference is even more surreal than it sounds. From across a cologne-scented hotel lobby, I watched tanned attorneys fresh off flights from the Caribbean mingle with ashen IRS agents who bring business cards from Kinko's because the agency won't pay to get them printed anymore. I listened to officials from the FBI and SEC lay out enforcement priorities as cryptocurrency investors and Russian bankers took notes. At lunch, I talked *Game of Thrones* with a Senate advisor, a government auditor and a Bahamian lawyer who later offered to set me up a shell corporation for $5,000.

The finance types were frosty during the day—an investor who appeared to be wearing monogrammed slacks wouldn't tell me his first name—but they loosened up at the happy hours. An offshore tax advisor bragged that he could take his clients' tax rates from 49 percent to 15 percent and complained that they were constantly pushing him to go lower. Another told me that most of his clients aren't trying to hide money from the government but from their second or third wives. "The first one raised their kids so they feel like she's entitled to something," he explained. "It's the trophy wives they want to lock out."

Jack Albertson is a government investigator—that's not his real name and he won't let me get more specific about his job description—

who has been coming to OffshoreAlert for years. When I ask him how this cops-and-robbers conflagration even exists, he tells me I'm thinking about it the wrong way. He, like all the other investigators here, knows that many of the lawyers who attend are hiding their clients' money sketchily or outright illegally. He even knows how they're doing it. The tactics for hiding money from tax authorities are not particularly sophisticated and have barely changed in the last 50 years. Set up a shell company and buy an appreciating asset—Iowa farmland, a London apartment, a New York pizzeria, something common enough that it won't attract attention.

Contrary to the *Catch Me If You Can* myth, Albertson said, solving financial crimes is not a cat-and-mouse game between cunning investigators and slippery con artists. Most of the time it is simply the blunt application of resources to a series of unimaginably tedious tasks. "Investigators can already crack almost any offshore account if they have enough time and money," he said. "The problem is that they only get that for a few cases a year."

Over the last four decades, the agencies responsible for investigating elite and white-collar crime—roughly speaking, the IRS, SEC, the Occupational Safety and Health Administration, the Environmental Protection Agency and the FBI—have seen their enforcement divisions starved into irrelevance. More than a third of the FBI investigators who patrol Wall Street were reassigned between 2001 and 2008. Enforcement funding at the IRS has fallen by 23 percent over the last decade. And, worst of all, every time a scandal exposes the government's inadequacy, Congress steps in to squeeze the regulators even harder.

The most instructive case of this deliberate stunting is the Consumer Product Safety Commission. Founded in 1972, the CPSC's job is to make sure the things you buy won't pierce, poison or burn you. In the 1980s, Ronald Reagan slashed its budget as part of his crusade against bureaucratic waste. In the 1990s, Clinton instructed the agency to produce more data as part of his push for government accountability. No matter which party was in power, every administration gave the CPSC more to do and less money to do it with. By 2007, it had shrunk from its initial 786 employees to just 420.

That same year, Mattel announced a recall of more than 1 million of its children's toys, which had been contaminated with lead paint. Despite the company's sophisticated international operations and billions in revenues, it had never bothered to inspect the Chinese sub-contractors. By then, the CPSC had fewer than 100 inspectors to monitor all imports to the United States. The Los Angeles–area ports where a chunk of the tainted toys arrived was overseen by a single part-time inspector.

Congress responded to the scandal by compounding the mistakes that had caused it. Lawmakers agreed to double the CPSC's budget and increase its staff, but also obligated the agency to carry out dozens of new activities, including the creation of a public database to track safety hazards for every single product sold in the U.S.

The new mandate swallowed up all the agency's new funding and more. Soon, the CPSC was dedicating nearly all of its time to lead abatement in children's toys, neglecting millions of products that posed far greater risks to children, like flammable blankets and dangerous table saws. The product database filled up with unconfirmed complaints and spammy comments. Mattel, meanwhile, faced no consequences for manufacturing the lead-tainted toys beyond a $2.3 million fine—roughly 0.006 percent of its net income. According to Rena Steinzor, the author of *Why Not Jail? Industrial Disasters, Corporate Malfeasance, and Government Inaction,* the same cycle has repeated itself across every form of elite deviance, from tax compliance to financial regulation to environmental protection. In 2010, following a series of tax-haven scandals, the IRS set up a "wealth squad" to investigate the ultra-rich—but only staffed it with enough agents to perform 36 audits in its first two years.

After the Enron-led avalanche of corporate bankruptcies in the early 2000s, Congress gave the SEC enough funding to hire 200 new auditing staff. At the same time, however, lawmakers obligated the agency to review the filings of every publicly traded U.S. financial firm every three years—a mandate far larger than the agency's new staffing levels. Then, after the financial crisis, it happened again: The Dodd-Frank Act tasked the SEC with monitoring even more companies and trillions of new assets while increasing its enforcement staff by less than 10 percent.

This cycle has left America's regulators with no choice but to engage in an increasingly desperate pantomime of white-collar law enforcement. On the outside, they report impressive performance statistics to avoid even more budget cuts. Behind the scenes, they've retreated to investigating only the defendants they know are guilty and the crimes they know where to find.

The primary beneficiaries of this shift are American elites. Rich people generate mountains of financial data. Millionaires can have over 100 bank accounts; billionaires' tax returns run to 800 pages long. For people who earn most of their income from working (i.e., almost everyone), the IRS has an automatic system that compares individuals' reports to the records submitted by their employers and banks. For the wealthy, who make much of their income from interest and investments, the agency has nothing to compare their reports against. The only way to tell if a rich person is cheating on their taxes is to sit down and go through them line by line.

"Let's say you get a tip that some billionaire is hiding a bunch of money offshore and not paying taxes on it," said Arthur VanDesande, who spent 25 years as a criminal investigator for the IRS. "And you manage to narrow the tax evasion down to 20 of his bank accounts. OK, now you have to prepare 20 subpoenas, get them signed by a judge and deliver them to the banks. But when you go to Bank of America, they say, 'We don't accept subpoenas at this location, you have to go to our authorized representative in Orlando.' So then you go to Orlando and you find out the money is linked to an offshore account. So then you have to write to the embassy . . ."

Due to the IRS's lean resources, VanDesande did most of this legwork himself. "You type your own shit, you make your own copies, you write every single affidavit. Sometimes you feel like, 'I'm a senior-level person with a college degree. Why am I calling Wells Fargo and sitting on hold for 45 minutes?'"

Only some of this drudgery can be outsourced to lower-level staffers. White-collar cases involve understanding arcane laws, absorbing thousands of pages of documents, traversing international jurisdictions and coordinating a vast array of agencies from the Secret Service to the USPS. They require investigators to be Jack Ryan, Magnum P.I. and Leslie Knope all at once. Even though auditing millionaires and billionaires is one of the most cost-effective government activities imaginable—an independent report estimated in 2014 that it yielded up to $4,545 in recovered revenue per *hour* of staff time—the IRS investigated the returns of just 3 percent of American millionaires in 2017.

In addition to reducing their caseload, America's white-collar enforcement agencies have started prioritizing crimes they can prosecute in bulk. In 2017, the Department of Justice took on 889 prosecutions for identity theft (which, according to a 2010 survey, is estimated to cost individuals $371 out-of-pocket) and just 24 for antitrust violations. According to a ProPublica report, 43 percent of all tax filers audited by the IRS earn less than $56,000 per year. Roughly one in six of the SEC's enforcement actions in fiscal 2019 were against financial firms for filing paperwork late—a sixfold increase since 2004.

Helen Richmond, a paralegal in a white-collar prosecutor's office (that's not her real name), said most of the defendants her office pursues are "either dumb or unlucky." She's worked on cases against money launderers who named stolen items on their wire transfers and fraudsters who sent emails with recipe-like details of their schemes. Criminals with even a scrap of sophistication, Richmond said, mostly avoid detection.

Another bargain-hunting strategy is to try cases in administrative proceedings rather than civil courts, an innovation that reduces hearings from months to hours. The downside, though, is that these cases largely play out in secret, resulting in fines rather than prison time, and don't compel defendants to testify, turn over evidence or admit guilt. In 2007, the SEC filed 60 percent of its settlements in civil courts. In 2015, that proportion was 17 percent. In 2014 and 2015, the agency didn't file a settlement against a single U.S. Wall Street firm.

According to former OSHA assistant secretary David Michaels, these strategies are designed to achieve the all-consuming yet unstated goal of every regulation agency in America: *Make yourself look more powerful than you are.* The best way to do this is to focus on the cases that will yield the maximum deterrence for the lowest cost. At OSHA, Michaels said, "we would issue press releases announcing waves of random inspections so employers would look at their hazards. We never told them we only planned to do a few inspections."

Similarly, the IRS has explicitly instructed agents to prioritize cases likely to generate headlines. (Ever wonder why so many B-list celebrities get busted for tax evasion?) Federal investigators go after media punching bags like Martin Shkreli, Martha Stewart and Fyre Festival scammer Billy McFarland to make the public think criminal prosecutions are routine. They're not: In a case-by-case analysis of the 216 alleged large-scale corporate frauds discovered between 1996 and 2004, researchers found that the media uncovered twice as many as the SEC.

And so, after decades of operating in survival mode, white-collar enforcement agencies are better at reporting success than producing it. In a 2016 study, a Georgetown University Law Center professor named Urska Velikonja discovered that while the SEC reported a steady rise in prosecutions between 2002 and 2014, most of the increase was statistical padding. When a trader was charged with fraud and then lost her license to trade securities, for example, the SEC logged the sanctions as two separate cases. When the agency was in danger of posting sluggish performance stats for the year, investigators filed dozens of slam-dunk cases in September to catch up. As Velikonja put it, "They're engaging in their own version of accounting fraud." (The SEC declined to comment.)

For the agents charged with cracking offshore tax schemes and protecting consumers from lead-painted Elmo collectibles, this charade is profoundly demoralizing. "We killed a generation of agents," VanDesande said. "If you investigate mom-and-pop grocery stores for 20 years, you lose the ability to do Bernie Madoffs."

VanDesande has spent months building cases only to have the DOJ toss them with little explanation. Richmond, the paralegal, tells me federal and state prosecutors have been playing hot potato with one of her cases for months because they can't justify an expensive prosecution for a fraud that adds up to the low six digits. During his first year on the job, Lewis Winters, an SEC examiner (and another government employee who couldn't use his real name), had an investigation of a shady CEO rejected by the agency's enforcement division. Even though he had found plenty of violations, the crimes just weren't . . . grand enough for the agency to pursue.

"It felt personal," Winters said. "Why did I spend three months examining this guy if enforcement just goes, 'meh'?"

3. DEFINING DEVIANCE DOWNWARD

Enron used to be considered the capstone to the golden age of white-collar prosecutions, a shining example of the system working like it's supposed to. Weeks after the company filed America's then-largest corporate bankruptcy, federal agents searched its headquarters and discovered a $63 billion game of three-card monte. Using an intricate network of off-the-books shell companies, Enron executives made loans look like income and debt look irrelevant. The year before the company collapsed, its leaders had falsified 96 percent of its net income and 105 percent of its cash flow.

Between 2002 and 2006, the FBI's Enron Task Force filed charges against more than 30 architects of Enron's fraud. Investigators discovered a "shred room" at the company's financial auditor, Arthur Andersen, and convicted the company of obstruction of justice. Four Merrill Lynch bankers were found guilty of helping Enron falsify its financial returns by purchasing three Nigerian barges. Task force agents convinced the company's chief financial officer to testify against his higher-ups by threatening to charge his wife with a felony. He flipped; they convicted her of a misdemeanor.

Eventually, after a five-year investigation, Enron founder Ken Lay and former CEO Jeffrey Skilling were convicted of securities fraud and a meal deal of lesser charges. Though Lay died at a rented mansion in Colorado shortly afterwards, Skilling got 24 years in prison. At the time, it was one of the longest white-collar sentences in U.S. history. Prosecutors called it a victory. Skilling's lawyers called it just the beginning.

As soon as the nation turned its attention elsewhere, Skilling's lawyers began quietly dismantling his sentence. They filed appeals objecting to the statutes used to convict him, the trial's Houston location and the questionnaires filled out by potential jurors. In 2013, citing the "extraordinary resources" it had spent prosecuting and defending Skilling's conviction, the Department of Justice agreed to cut ten years off Skilling's sentence if he promised not to file any more appeals. He was released in February 2019 after serving less than half his original sentence.

The rest of the FBI's victory has crumbled under the same blitzkrieg of high-priced lawyering. The Supreme Court overturned Arthur Andersen's conviction in 2005. The convictions of three of the Merrill Lynch bankers were vacated after they convinced an appeals court that they were merely trying to "solidify business relationships" rather than acting for personal gain. In the end, just 18 people served prison sentences (by comparison, more than 500 served time for the savings and loan crisis of the 1980s and early 1990s). Fourteen of them served fewer than four years. Andrew Fastow, the mastermind of Enron's network of shell companies, now makes his living lecturing business school students and fraud investigators about how he did it.

Nearly every high-profile corporate scandal has the same overlooked epilogue. The wealthy have always attempted to spend their way to lighter sentences, but in the last two decades, the American judicial system has become increasingly willing to let them.

"We've seen a concerted effort to define deviance downward," said Paul Leighton, a professor at Eastern Michigan University and the coauthor of *The Rich Get Richer and the Poor Get Prison*. "We've made felonies into misdemeanors, misdemeanors into torts, and torts into regulatory offenses."

Honest services fraud, for example, is the subsection of mail and wire fraud that prohibits companies from lying to customers to get their business and CEOs from lying to investors after they've already been hired. Think of a mechanic telling you that your perfectly functional transmission is busted, then telling you it will cost $2,000 to fix it. He hasn't defrauded you exactly—he really will replace your transmission—but he used his position of authority to scam you into paying for something you didn't need.

Since 1909, prosecutors have used the honest services fraud provision to go after companies that lie to boost their stock price and politicians who give golfing buddies lucrative procurement contracts. District court judge Jed Rakoff, a former white-collar prosecutor, once referred to the statute as "our Colt 45, our Louisville Slugger, our Cuisinart."

But over the last three decades, the Supreme Court has taken the law apart piece by piece. In 1987, the Rehnquist Court ruled that the statute should never have been used to protect the so-called "right to honest services." In 2010, the court restricted its application to public-sector bribery and kickbacks. From now on, the lying mechanic is breaking the law only if *someone else is paying him* to scam you.

Based on that ruling, several white-collar criminals—including, wait for it, Jeffrey Skilling—had their sentences or convictions vacated. This year, two former Chris Christie underlings will tell the Supreme Court that orchestrating the "Bridgegate" conspiracy, in which they deliberately orchestrated traffic jams to get revenge on a Democratic mayor, is no longer illegal under the new, narrowed definition. If the Supreme Court agrees, the law will get even weaker.*

Other white-collar statutes have suffered the same slow strangulation. In 2006, a district court judge reaffirmed the right of companies to pay the legal fees of their executives, effectively giving every C-suite defendant the same deep pockets as their corporate employer. Since 1996, the Supreme Court has consistently blocked plaintiffs from receiving punitive damages, arguing that large punishments deprive corporations of their due process rights. In 2016, the court ruled that federal bribery law only applies to politicians who traded official acts for personal benefit—the kind of immediate, explicit kickback that rarely happens outside of corporate HR training videos.

"Criminal law used to be more closely aligned to our moral intuitions," said Will Thomas, a University of Michigan professor who studies corporate liability. "We still talk about it like it's a guiding moral force, but it's a much more administrative process now."

Today, Thomas explained, judges are more willing to disregard the consequences of their rulings (like, say, an Enron-scale fraud going unpunished) in favor of resolving obscure procedural ambiguities. In 2017, for instance, a case against New York financier Benjamin Wey was dismissed after he successfully argued that the search warrant used to gather evidence against him was overly broad and vaguely worded.

---

* The Supreme Court did overturn the convictions of the Chris Christie underlings later in 2020.

The confounding thing about these challenges is that they often highlight real weaknesses in the criminal justice system. American law is a contradictory jungle of century-old statutes and arbitrary definitions. Lying to government investigators, for example, is prohibited by at least 215 separate laws, each with their own standard of proof. *Mens rea,* the concept of "guilty mind" central to establishing criminal liability, has more than 100 definitions across various statutes.

So of *course* wealthy defendants win cases by arguing that fraud statutes and insider trading rules are poorly written. They are. But so are the rest of the laws. (Numerous state anti-gang statutes, for example, define "gang" so imprecisely that they could apply to most sororities.) The only difference is that white-collar defendants have the ability to dispute every step of the process used to convict them—and a judicial system all too happy to oblige.

One of the most conspicuous aspects of white-collar cases is the doting, near-veterinary care with which judges try to prevent defendants from facing harsh punishment. In 2014, a Colorado judge ruled that two farm owners whose tainted cantaloupes caused a listeria outbreak that killed 33 people couldn't be sent to prison because it would interfere with their ability to earn income for their families. As he announced a sentence of five years' probation, the judge explained, "I must deliver both justice and mercy."

According to a study by the Federal Judicial Center, four out of five judges in federal courts (where the vast majority of white-collar cases are decided) are white. A 2010 survey found that they have an average age of nearly 70. Their base salary is $210,000 per year.

It is, as one of those high-priced lawyers might say, *improbable* that these demographic and economic facts exert no influence whatsoever on judges' rulings. In a 2012 review of sentencing data in Florida, researchers found that "high-status" white-collar criminals, such as doctors scamming Medicaid, were 98.7 percent less likely to receive prison terms than welfare fraudsters. A 2015 study found that judges showed increasing mercy as fraud offenders moved up the income scale: Criminals who stole more than $400 million got sentences that were less than half of the minimum recommended by federal guidelines. Criminals who stole $5,000 or less served sentences well over the minimum.

"When you zoom out, you see all the ways that bias accumulates throughout the system," said Justin Levinson, a University of Hawaii professor and coeditor of *Implicit Racial Bias across the Law*. Sentencing guidelines prescribe lighter punishments for first-time offenders and criminals who can afford to pay restitution. Evidence rules make it nearly impossible to seize records or computers from corporations. Jury selection weeds out the poor, the less educated and minorities.

And then there's the matter of criminal liability. For many low-level crimes, prosecutors have to prove that a defendant *should have known* a crime was taking place. If a renter deals drugs out of her apartment, her landlord can be prosecuted. If you loan your friend your car and he commits a murder while driving it, you can be charged with murder, too. For executive-level crimes, however, the bar of criminal liability is set impossibly high: Prosecutors have to prove that defendants knew their actions were illegal and did them anyway. This myopic focus on intent means that white-collar trials often come down to the question of whether the defendant was the kind of person who would commit a criminal act.

John Lauro, an attorney who has represented healthcare and

financial executives, said he always emphasizes the complexity of white-collar crimes to the jury—*financial disclosures are so technical! How could my client possibly know that stock wasn't going to pan out?* He also plays up the upstanding-citizen angle. The first thing he does when he lands a new client, he said, is visit their home and meet their family.

"I bring in things like their marriage, their kids and whether they coach Little League," he said. "The prosecution always wants to dehumanize them. They call my client 'the defendant.' I'll call him by his first name until a judge tells me to stop."

The only way to get around this, said Sarah Larkin, a securities fraud prosecutor in Manhattan (she couldn't speak on the record, so that's not her real name), is to make every crime seem as simple as possible. It's lying, it's cheating, it's stealing. She structures every trial like a crash course, spending days explaining how the stock market works and what acronyms like SEC, CDO and GAAP stand for. Before she can convince jurors that the defendant lied on a financial statement, she has to do a weeklong *Big Short* interlude to teach them what a financial statement even is—without the help of Margot Robbie in a bathtub.

"And after all that," she said, "you still have to make the case for why this person who looks very upper-middle-class and has a family sitting in the back row should be branded a criminal. It's a heavy lift."

The near impossibility of establishing white-collar defendants' motives combines with the high standard of reasonable doubt to create a paradox. Most Americans have a visceral aversion to greedy executives *in general*. Introduce them to a single banker and a specific crime, however, and their moral outrage often melts away. As Sam Buell, a Duke University law professor, told me: "Put people on a jury and they'll say, 'Gee, it seems like this guy was doing his job, so I don't think it was a crime.'"

Take the case of Brian Stoker, a Citigroup employee who was charged in 2011 with marketing risky investments (one trader called them "dogsh!t" in an internal communication) as safe bets. According to the SEC, the bank made $160 million while investors lost $700 million. In his closing argument, Stoker's attorney showed the jury an illustration from a Where's Waldo? book. Their client was a nobody, he suggested, a scapegoat for the culture of high-stakes gambling that had taken over the entire financial sector. Why make him a patsy when everyone else was doing the same thing?

The jury declared Stoker not guilty. But in the same envelope as their decision, they included a handwritten note. "This verdict," it read, "should not deter the S.E.C. from continuing to investigate the financial industry." In other words: Keep trying to lock up greedy bankers. Just not this one.

And this is it, the Rosetta stone for understanding why judges are so comfortable explaining away the misconduct of corporate executives; why Congress never strengthened the castrated white-collar statutes; why so few pharmaceutical executives have been imprisoned for the opioid crisis and only a single banker went to prison for the financial crash. American law is incapable of prosecuting crimes in which elites use their legitimate power for nefarious ends.

"The way businesses harm people is the same way they interact with them normally," Albertson said. Banks collect debts and foreclose on homes every day. Banks give out home loans every day. When they entice customers into unaffordable mortgages or foreclose on borrowers tricked into signing loans they can't afford, the courts can't tell the difference.

This insight also explains why the legal system applies the opposite logic to organizations run by the rich and organizations run by the poor. Teenage gang members who argue that they committed crimes due to the culture of the Crips or the Latin Kings receive harsher sentences—stealing money for yourself is bad; stealing money for a criminal organization is worse. Corporate defendants who claim they committed crimes due to the internal culture of Goldman Sachs or HSBC, on the other hand, get *lighter* sentences—how could an individual possibly be held accountable for something everyone else was doing?

And so, as they lose the ability to prosecute high-level crimes and elite offenders, many of America's criminal justice institutions have simply stopped trying. Of the 649 companies prosecuted by the Department of Justice since 2015, only eight were convicted in court. The rest either took settlements or negotiated themselves a deferred prosecution or non-prosecution agreement.

These arrangements, like so many other aspects of America's white-collar enforcement apparatus, represent the cynical perversion of a benign idea. Deferred prosecution agreements were created in the 1930s to allow first-time juvenile offenders to avoid jail time if they followed probation rules and didn't reoffend. In the early 1990s, prosecutors began extending the principle to corporations: If you agree to investigate your own crimes, turn over evidence against your employees and change your internal policies, we won't take you to court.

Since then, deferred prosecutions have become one of the primary engines of American impunity. They don't require companies to explicitly admit guilt and don't apply steeper punishments to repeat offenders. While courts often appoint independent monitors to make sure corporations comply with the terms of their probation, these reports aren't released to the public. Since 1999, only three companies have ever been prosecuted for violating the terms of their agreements.

"Criminal law isn't just about deterrence, it's about moral education," said John Coffee, the director of the Center on Corporate Governance at Columbia Law School. "You show the public that a crime occurred and how terrible its impact was. We're missing that catharsis now."

4. THE CONUNDRUM OF OVERDUE CONSEQUENCES

For multinational grocery chains, self-checkout kiosks are a no-brainer. They save space, cut costs and speed up lines for shoppers. There is, however, one downside: Allowing customers to scan their own groceries dramatically increases the proportion of people who shoplift.

What self-checkout kiosks provide, researchers have found, is plausible deniability. If a security guard spots you slipping a pack of Tic Tacs into your pocket, there's no way to cast yourself as anything but a thief. If he catches you keying in a $10 bag of trail mix as a $2 bag of lentils, you can call it a mistake—*oops, I must have typed in the wrong code!* Perpetrators, especially middle-class white ones, know that if they get caught, everyone from the store manager to a small-claims court judge is likely to give them the benefit of the doubt. Self-checkouts turn shoppers into shoplifters by providing them with an opportunity to steal and a ready-made excuse to get away with it.

Nearly all criminological research indicates that crime rates depend more on environments and incentives than the intrinsic morality of offenders. It's why shootings spike on hot days and drivers speed up on wider streets. People aren't good or bad; they drift into good or bad behavior when one or the other is rewarded.

You can see where I'm going with this. Since the 1980s, Wall Street, Congress and the courts have systematically encouraged American elites to commit more and larger graft. "Corporate culture warps people," said Mihailis Diamantis, a University of Iowa professor who specializes in corporate crime. "They've been placed in institutions that facilitate lawbreaking and predispose them to break the rules." Since 2009, the percentage of employees at large companies who report that they've been pressured to commit ethical breaches has doubled. In a 2015 study, more than half the auditors for the country's largest companies said they had been asked to falsify internal audit reports. In Ernst & Young's 2016 Global Fraud Survey, 32 percent of American managers said they were comfortable behaving unethically to meet financial targets.

And just like those shoppers standing in front of that unmanned kiosk, the scriptures of corporate America discourage white-collar criminals from reckoning with the reality of their crimes. Larkin, the Manhattan prosecutor, said that when she used to prosecute murderers, they would strike a plea deal and then immediately open up—here's why I stabbed him, here's where I hid the knife. Once she switched to elite criminals, she was floored by their utter refusal to take responsibility. "They minimize and make excuses," she said. "They believe in their own brilliance. They keep saying what they did wasn't really wrong."

Jack Blum, a former staff attorney for the U.S. Senate, calls this impunity "the most urgent issue in America." In Russia and Ukraine, as government capacity deteriorated during the 2000s, oligarchs increased spending on bribes, lobbying and parallel systems of power—their own private security forces, their pet media institutions. The same thing is already happening here: According to a 2016 analysis, political lobbying and regulatory maneuvering have eclipsed research and development as the primary reason for rising corporate profits. In a country of declawed regulators and untouchable executives, dishonest companies will increasingly drive out honest ones.

So how do we stop this? The obvious temptation is to bring back the glory days of white-collar prosecutions, to lengthen the sentences for CEO dirtbags and finally arrest the bankers that got us into the financial crisis. It felt *good* to hear that Martin Shkreli cried at his sentencing hearing, damn it. America deserves the catharsis of overdue consequences.

But retribution has been the government's approach for decades. Every corporate scandal since the savings and loan crisis has produced a wave of prosecutions and a sprint to strengthen white-collar punishments. The maximum penalty for the most common securities fraud charge is already 20 years in prison and a $5 million fine. Bernie Madoff is 120-odd months into a 150-year sentence.* It might feel good, but there's little indication that handing out life sentences for the bankers who caused the last crash will prevent the next one.

That's because criminologists have consistently found that increasing the likelihood of punishment works better than increasing its severity. In a study of wastewater discharge from chemical plants, researchers noticed that managers who received large but inconsistent fines actually started emitting *more* toxic chemicals. The harsh penalty may have made them resentful for being singled out for

* Bernard Madoff died in prison in 2021.

something everyone else was doing—and the low probability of getting caught twice encouraged them to increase their lawbreaking to catch up with their competitors. Locking up one corporate criminal out of a million might make the other 999,999 feel even more entitled and invincible than they do now.

And yet elites, like everyone else, do change their behavior after experiencing immediate, reliable consequences. Three independent studies have found that when the wealthy get their taxes audited, they cheat less the following year. In 2008, Norway offered a "tax amnesty" to its richest residents. If they reported their overseas wealth and paid taxes on their hidden income, they would be immune from prosecution. For the next four years, the targets of the amnesty reported a 60 percent increase in their net worth and paid 30 percent more in taxes.

"If you follow a company over its life cycle, studies have found that most of them engage in some kind of lawbreaking and almost all of them reoffend," said Sally Simpson, a University of Maryland professor and coauthor of *Understanding White-Collar Crime: An Opportunity Perspective*. "The way you get deterrence is by showing them they're being watched."

In a 2016 review of dozens of studies on corporate crime and deterrence, researchers found that nearly every individual strategy for punishing companies and executives had little to no deterrent effect on its own. The only thing that consistently worked was to combine them—warnings from government agencies, surveillance of the worst actors, harsher punishments for repeat offenders and, yes, at the top of the ladder, criminal prosecutions for corporations that refused to shape up.

This is hardly some exotic, untested concept. When it comes

to every other form of crime, law enforcement agencies are perfectly comfortable cracking down on offenses at every level. This is the country that invented three-strikes laws and "broken windows" policing. When it comes to street gangs and drug distribution networks, the criminal justice system has no problem simplifying complex criminal liability questions into four simple words: *You should have known.*

"The law is just a way of saying that an immoral act is something we're not going to tolerate anymore," said Robinson. "It's up to us to decide what's a crime and what isn't. We do it all the time."

We've just never done it for the immorality of elites.

ORIGINALLY PUBLISHED IN

***HUFFPOST HIGHLINE*, FEBRUARY 2020**

PICTURESQUE CALIFORNIA CONCEALS A CRISIS OF MISSING INDIGENOUS WOMEN

BY BRANDI MORIN

On September 7, 2018, Angela McConnell, 26, and Michael Bingham, 31, were found shot to death in a wooded area in Shasta Lake, a small town in northern California. The police didn't tell Tammy Carpenter, McConnell's mother, that her daughter had been murdered. Instead, the Hoopa Valley tribal member found out from one of Bingham's relatives.

When Carpenter drove to the police station to find out if her daughter was really dead, the officer who greeted her asked if she knew that her daughter was on drugs and living as a transient.

"I thought, '*What the hell is this?*'" says Carpenter. "I told him, 'You need to go do your job and find out who murdered my daughter!'"

But more than three years later, the murders of McConnell and Bingham remain unsolved. Police told Carpenter they didn't have enough evidence to move forward; she says the murder scene was left unsecured.

"Like all these missing or murdered Native women, she doesn't matter," says Carpenter. "They won't solve her case because she's just another Indian."

Angela McConnell is one of thousands of missing and murdered Indigenous women and girls (MMIWG) across the United States. For many years, authorities overlooked the crisis but now families and community members are demanding justice for crimes that they say stem from centuries of oppression.

The National Crime Information Center, a federal agency, has documented more than 5,000 cases of missing Indigenous women. Experts say the real number is likely higher. Eighty-four percent of Indigenous women experience some form of violence during their lifetimes while those living on reservations are killed at 10 times the national murder rate.

"It's racism," says Annita Lucchesi, founder and executive director of the California-based Sovereign Bodies Institute (SBI), which researches and addresses violence against Indigenous women. "Authorities don't protect Indigenous women."

Violence against Indigenous people has a long history, going back to the early days of colonization and extending to include slavery, land seizure, the forced removal of children from their families, and multiple massacres. In California, according to one of SBI's reports, "historians estimate that as many as two out of three California Indians were killed in the two years following the [1849] discovery of gold."

Native American families continue to contend with this "bloody legacy," as the report calls it. Their daughters, sisters, and mothers are vulnerable, says Lucchesi, and predators know it. Police are less likely to investigate missing Indigenous women, known perpetrators are less likely to be prosecuted or convicted, and the media is less likely to cover MMIWG cases with the same alarm as those of missing white women.

"A lot of times the places they go missing from are extremely rural," Lucchesi says. "There's a lack of services, a lack of transportation, and a lack of opportunity."

That's especially true in California. The state as a whole has the highest Native American population in the United States—more than 750,000 people belonging to nearly 200 tribes, many of them living in the sprawling metropolises of Los Angeles and San Francisco. Yet those urban areas appear to be far less dangerous for Indigenous women than rural northern California, with its soaring redwood forests and rugged shoreline. There, at least 107 women have been murdered or gone missing since 1900, twice the number as in the Bay Area, where the Indigenous population is three times the size.

—

HOOPA VALLEY TRIBE MEMBER ARNOLD DAVIS III, 32, MISSES HIS MOTHER, even though he doesn't remember her. "I'm told she was outgoing, kind of like a free bird," he says. "Everybody always said that she was beautiful."

Davis was two years old when Andrea Jerri White, also known as "Chic," disappeared on July 31, 1991, near the town of Blue Lake in Humboldt County.

"Nobody talked about it in the house growing up. . . . I think it hurt my grandma a lot and she didn't know how to break it to us kids," says Davis, sitting with his hands folded at a long table made of redwood at the SBI office in Eureka.

White—half Hoopa, half Yurok—was last seen near an exit on California Highway 299. According to police reports, she was hitchhiking home to the Hoopa Valley after attending a court hearing in Eureka. The 22-year-old had been charged with a DUI after a car accident a few weeks earlier. The state took custody of her four children, who had been in the car, and placed them with their grandmother.

"She was fighting to get us back. That's why she hitchhiked to court," Davis says.

His loss didn't really hit him until he was working on a Mother's Day project when he was in elementary school. "You had to have your mom's name and, for each letter that's in it, write something that describes your parent," he says. "I realized I didn't even know my mom's name."

Davis believes the police could be responsible for his mother's disappearance, but he has no evidence. Not long before White disappeared, she told her aunt, Donna White, that she'd been raped by local police officers. "She said she had been drinking and they picked her up for intoxication," says Donna. "But they stopped, and they raped her. They still put her in jail, too."

Donna took her niece to a rape crisis center in Eureka and then to the district attorney's office. But it proved too much for White: "The district attorney was talking to her and then she just shut down and didn't want to talk about anything."

Police have no leads on the disappearance, now relegated to the cold case file. Humboldt County sheriff William Honsal says foul play is highly suspected.

—

IN THE RURAL COMMUNITY OF ARCATA, NINE MILES NORTH OF EUREKA, LUC-chesi, 29, frantically searches for an outlet to charge her cell phone. She is fielding calls nearly nonstop as she tries to coordinate efforts to help MMIWG survivors.

"Every day I'm buying groceries or paying rent or, you know, helping somebody escape an abuser," she says. "We have a line item in our budget called grandmas because we have so many grandmas that are raising grandkids because the parent is missing or murdered, and they don't have the money for that, much less the help and the energy."

Lucchesi's personal and academic background makes her uniquely suited for this work. A domestic and sexual abuse survivor, she embraced her Cheyenne heritage in college and is pursuing a PhD in Indigenous cartography and geography. When she became aware of the lack of data on violence and death among Indigenous women and girls, she created a database of her own. The Sovereign Bodies Institute database has since become known as one of the most comprehensive collections of information about MMIWG in North America.

"This is not a mystery to be solved. We already know what's happening. And we know how to fix it. We know where the problems are, and we have the minds capable of fixing it," she says. "It's something that needs aunties at a kitchen table with the power to do what's best for the community."

—

RONNIE AND LYDIA HOSTLER HAVEN'T SEEN THEIR GRANDDAUGHTER, KHADIJAH Britton, in four years. The 23-year-old former high school basketball player, a Wailaki member of the Round Valley Indian Tribes, was kidnapped at gunpoint in Covelo on February 7, 2018.

Britton's family and friends have plastered missing person posters throughout Covelo and nearby communities, backed an aggressive social media campaign for information on her whereabouts, and attended countless rallies carrying her banner.

It's exhausting work. But Ronnie, 78, and Lydia, 73, say they'll never give up looking for Khadijah.

"We need to do this, not only for our granddaughter, but I don't want to see so many other families going through what we're going through," says Ronnie.

Nearly half of the MMIWG cases, including Britton's, in northern California are due to domestic or intimate partner violence. Before she vanished, she'd filed a restraining order against her ex-boyfriend, Negie Fallis.

"He started getting violent with her," says Ronnie. "She got into some bad things, you know. I hoped she'd come out of it, but she didn't."

A few days after Britton took out the restraining order, Fallis caught up with her at a party on the outskirts of Covelo. According to eyewitnesses, he brandished a small derringer pistol and forced Britton to leave with him and another woman in a black Mercedes sedan.

Three days later Britton's stepmother filed a missing person report. At first, police didn't suspect foul play. Now Fallis, who has been arrested on multiple unrelated charges, is their main "person of interest."

Ronnie Hostler believes police should have acted sooner. But resources are strained in Covelo. The tribal and federal police have limited jurisdiction because all tribes in California fall under Public Law 280, which requires the state—rather than tribal or federal law enforcement—to prosecute most crimes that occur on a reservation.

And crime rates are high. Covelo sits within the Emerald Triangle, a trio of counties—Humboldt, Mendocino, and Trinity—where the largest marijuana producers in the United States operate. Cannabis is legal in California but criminal organizations, including cartels, remain active.

"I think that this contributes to some of the violence that is happening there," says Round Valley Tribal Police chief Carlos Rabano Sr.

While family members wait for answers, authorities have taken some action in the Britton case.

Fallis remains in prison. An anonymous tip line has been established. There is an $85,000 reward for information that leads to a conviction in her case, plus a $10,000 reward from the FBI. Family members, volunteers, and police have searched for miles in and around Covelo for any signs of Britton, but they've found nothing.

Fallis "is my leading suspect," says Mendocino County sheriff Matt Kendall. "But at this point, we need people who are going to speak. We need people who are going to talk about what they know. We've had numerous instances after interviewing witnesses who have changed their stories two, three times. And that puts some real hurdles in front of the district attorney when it comes to charging these cases."

Kendall says the relationship between law enforcement and Indian Country is broken. Although he is not Indigenous, Kendall grew up in Covelo, which is partially located on the Round Valley Indian Reservation. When his deputies ask why Native Americans don't trust the police, he offers this answer: "If North Korea came over here tomorrow and attacked us and killed eight out of 10 of everybody, and then they told you that you're going to leave here and you're only going to speak North Korean, what would you tell your kids about those guys?"

—

ABBY ABINANTI, CHIEF JUDGE OF THE YUROK TRIBAL COURT IN KLAMATH, 200 miles to the north, addresses the lack of trust more bluntly. "The Yurok word for policemen translates to 'men who steal children,'" she says. "The first time we ever met them was when they came and stole children as indentured slaves or for the boarding schools. So, you have a natural resistance on our part."

The first Native American to be admitted to the California bar, 74-year-old Abinanti served as a San Francisco Superior Court commissioner for two decades before returning home to Yurok territories 25 years ago. She says her approach to justice is grounded in Yurok culture, incorporating concepts of community, culture, fairness, and responsibility—not degradation and punishment.

"I know them, I know their families," she says. "I know what they've been through."

Abinanti works with a team of 47 at the tribal court to develop ways to help Yurok and other people coming through the court resolve disputes, reunite families with children, and rehabilitate those who commit crimes on their territory. The MMIWG crisis is a big focus of her work.

"The numbers show that clearly it's a crisis and that [Indigenous women] are invisible and that the issues have not been addressed. And for me, this is our responsibility," says Abinanti, who cowrote recent SBI reports on the crisis. "These are our people. They need to come home."

Blythe George, a member of the Yurok Tribe who works with Abinanti and has also helped write SBI reports, expressed anger over the lack of attention paid to MMIWG by mainstream society. "This is not new," she says. "It's as old as anything else we've dealt with." But, she continues, "for a long time it wasn't important enough to keep track. It's hard not to see the complete lack of justice at times and not wonder, if these were white women, would it be different?"

—

FOR FAMILY MEMBERS, KNOWING THAT OTHERS ARE SHINING A LIGHT ON MMIWG cases can be incredibly powerful.

Thirty years ago, Christina Lastra's world was shattered when she received news that the body of her mother, Alicia Lara, had been found in the passenger seat of her car at the bottom of a dumpster in Weitchpec near the Hoopa reservation.

"The last time I saw my mother, she was bent over her camping gear and she looked so beautiful. Her hair was up in a bunch," remembers Lastra, wiping tears from her cheeks. "I said, 'Mom, I'm going to work now.' And she said, 'okay, Mihai,' which is 'my daughter.'" They hugged, said "I love you," and that was it.

Police declared Lara's death an accident. But a year later the local coroner told Lastra that her mother's autopsy report showed Lara, who was Tarahumara, had been murdered.

An anonymous witness told Lastra that someone had seen her mother bleeding in the passenger seat of her car. Lastra is convinced that the driver was an accomplice of her stepfather, a white illegal marijuana grower who died two months later.

Lastra believes her mother's death was classified as an accident because authorities did not deem her case worth investigating. "It's unfortunate that we [Indigenous] are treated as if we don't count."

When Lastra met Lucchesi a few years ago and learned about SBI's work, she felt a sense of relief knowing there was a network of support for missing and murdered Indigenous women like her mother. "I felt my mom was being seen, recognized, honored," she says. "And that was the day that I started to heal from the murder of my mom."

—

LAWMAKERS ARE BEGINNING TO ADDRESS THE CRISIS. IN CALIFORNIA, JAMES Ramos, the first California Native American to be elected to the state assembly, sponsored legislation aimed at untangling criminal jurisdiction among law enforcement agencies and tribal governments, improving public safety on tribal lands, and researching the impediments to reporting and identifying cases of missing and murdered Native Americans in California, particularly women and girls. The legislation was approved in 2021.

Once the research is done, "we can strengthen laws where we need to strengthen them," says Ramos, a member of the Serrano and Cahuilla tribes. "Our role is to continue to heighten the awareness of what truly is happening in Indian Country, in the state of California, and across the nation."

Nationally, Secretary of the Interior Deb Haaland created a new Missing and Murdered Unit (MMU) within the Bureau of Indian Affairs Office of Justice Services, which will help put "the full weight of the federal government into investigating these cases." A task force on Missing and Murdered American Indians and Alaska Natives called Operation Lady Justice was formed in 2019 to pursue outstanding cases as well.

"Far too often, murders and missing persons cases in Indian country go unsolved and unaddressed, leaving families and communities devastated," said Haaland in a statement announcing the new unit. "The new MMU unit will provide the resources and leadership to prioritize these cases and coordinate resources to hold people accountable, keep our communities safe, and provide closure for families."

Change isn't going to happen overnight, says Abinanti. Other issues need to be addressed, such as the many vulnerable Indigenous children in foster care and strengthening Indigenous communities. "We need to make it better for our people at home so they can be at home."

There's been too much suffering, she says: "The crying, the children [are] crying for their moms who will never come back, and they don't stop crying their whole lives."

Eventually, Abinanti hopes to create a prototype for all Indian Country on how to find murdered and missing Indigenous women and girls and prosecute the perpetrators.

"I want a private investigator. I want those cold cases looked at. I want people interviewed. I want DNA testing. I want search dogs," she says. "I need to fix this now."

Meanwhile, Tammy Carpenter is tormented by a constant cloud of sorrow that's followed her since her only daughter was murdered. She struggles with getting out of bed every day.

"It's hard as a mother to try to move forward. I'll never plan a wedding or be a grandma for her kids," she says.

Despite her pain, she's determined to advocate for justice for McConnell and all victims of the MMIWG crisis.

"I'm her voice now. And I have to talk. . . . She didn't deserve to die."

ORIGINALLY PUBLISHED IN

***NATIONAL GEOGRAPHIC*, MARCH 2022**

HOW THE ATLANTA SPA SHOOTINGS—THE VICTIMS, THE SURVIVORS—TELL A STORY OF AMERICA

BY MAY JEONG

I. LYON

On the afternoon of March 16, 2021, Marcus Lyon and his girlfriend dropped off their son at day care and went out for a late lunch not far from Sixes, a suburb of Atlanta that takes its name from a collection point on the Trail of Tears. The seafood restaurant where they headed stood just south of Larry McDonald Memorial Highway, named for the Georgia politician who served as president of the John Birch Society, a Cold War–era group that viewed the civil rights movement as a communist plot.

Lyon, 31, had been delivering for FedEx since November, but the past month, he'd been suffering from lower back and shoulder pain. He earned $130 a day plus a dollar for every delivery after the 110th, and no benefits. He was faster than most of his colleagues, once making 25 deliveries in an hour, but the load was wearing him out, plus their four-year-old had been crawling into bed with them lately.

Around 4 p.m., Lyon, off work that day, dropped his girlfriend at the bar where she worked and began driving along his delivery route, which, as usual, took him by Young's Asian Massage in Acworth. Just before 5 p.m., he decided to go in. Inside, 44-year-old Daoyou Feng greeted Lyon and asked if he wanted "one or two girls." Lyon said one would do and paid Feng $120 in cash. She took him down a narrow hallway that ran the length of the parlor to a room on the left. Lyon took off his clothes, pulled a towel over himself, and laid facedown on the massage table.

Feng entered and began massaging Lyon's neck. No more than a few minutes had passed when they heard a gunshot. Then a second shot rang out. Lyon dove behind the table, still undressed. Feng opened the door, and a third gunshot hit her in the head.

Lyon remained hidden, waiting until the shots stopped and a doorbell quieted before he reached for his trousers and shoes. He rushed out to his car, grabbed his 9-mm pistol, purchased at a pawnshop two weeks prior, and ran back inside. Three women, all spa workers, stood wailing. They asked Lyon to call the police and rushed out.

Lyon saw another customer, Elcias Hernandez-Ortiz, walking around with blood dripping from his head. That afternoon, Hernandez-Ortiz had stopped by the strip mall, as he did every other week, to wire money to his family in Guatemala, where it supported five people. He'd been waiting for a masseuse when the shooting started.

When the killer opened the door to his room, Hernandez-Ortiz got on his knees, put his hands up, and asked to be spared. "Please don't shoot me. I haven't done anything. Please don't shoot, please don't shoot. . . ." A bullet entered his face between his nose and his left eye. Because Hernandez-Ortiz had been looking up, the bullet traveled through his nasal cavity rather than his brain, down his throat, lodging itself in his abdomen. After the killer left, Hernandez-Ortiz found refuge in the spa bathroom. He waited there for help.

Lyon called 911 and began narrating what he was seeing to the operator. Feng was dead. So was Delaina Yaun González, a 33-year-old Waffle House server who'd come to the spa with her husband, Mario González, who is from Mexico and who worked as a landscaper. Mario could only take time off in inclement weather, which was why the couple was visiting the spa as the rain began that day. Paul Andre Michels, a 54-year-old handyman there to check out a pipe, whom the workers called *lao zhang*, or "dear mister," was dead too.

Lyon could see one person still breathing. It was the owner, Xiaojie Tan. The 911 operator asked Lyon if he could administer chest compressions. Lyon, who used to be a lifeguard, declined. A few years prior, during a shooting at a FedEx warehouse in a different Atlanta suburb, a 19-year-old worker had shot six colleagues before killing himself. A FedEx employee, a certified emergency medical technician, tucked an injured security guard's organs back into his body. Later, the security guard, who had developed complications from the shooting, tried to sue his colleague, Lyon recalled. "People are crazy like that," he said. (The security guard, Christopher Sparkman, had sued FedEx, not his colleague.) A few minutes later, three police officers arrived. Lyon watched as they dragged González out in handcuffs and put him in a police car, where he was mistakenly detained for hours.

Back home that night, Lyon felt "weird" sleeping next to his girlfriend and his son—he had been too close to death to lie among the living—so he got up and went to the sofa. Three days later, Lyon was back at work, but the dull thud of boxes hitting pavement reminded him too much of gunshots, and by month's end he had quit.

"I am not going to let that happen again," Lyon told me when we met at a Dunkin' Donuts in Sixes, 10 miles north of Young's. His 9-mm goes everywhere with him now, even to bed.

STONE MOUNTAIN

On the other side of Atlanta, 16 miles due east, is Stone Mountain, then as now a Native American holy site. In 1945, Ku Klux Klan members climbed it to carve out a cross, stretching 300 feet across the mountain face. The men lit it on fire, a Pharos visible 60 miles away, according to historian Kevin M. Kruse in *White Flight*.

Upon this sacred stone face, sculptor and KKK sympathizer Gutzon Borglum had begun exerting his will—a bas-relief of Jefferson Davis, Robert E. Lee, and Stonewall Jackson—but left before finishing, moving on to his opus, Mount Rushmore. Stone Mountain is "the largest shrine to white supremacy in the history of the world," as Richard Rose of the National Association for the Advancement of Colored People has said. Stone Mountain is also among the most visited sites in the state. On any given weekend, families host cookouts, play mini golf, and line up blithely for leisure activities against the backdrop of the three horsemen.

II. FENG

When Daoyou Feng was 14 or 15, or maybe 16—accounts vary—she left home, a village near Zhanjiang prefecture in China, and moved

260 miles east to Guangzhou city, near Hong Kong, where she found work at a toy factory. Feng's family was desperately poor and relied on Feng and her older brother Daoqun, who left home when Feng was three or four to work at a rubber tree farm, where he made the equivalent of $5 a month. Another brother, Daoxian, whose foot was debilitated in a childhood injury, supported himself by farming. Her sister, Mei, also sent away to find work in the city, had eloped with a factory worker. And so the weight of filial duty fell on the shoulders of young Feng.

After working at various factories in Guangzhou and Shenzen, Feng went to work in Shanghai. She told her family she gave facials. The year she turned 38, Feng returned to her village to find a husband. The search was unsuccessful. "They tried to introduce men to her," Daoqun recalls. "But she wouldn't even take a look. 'This is not okay.' 'That is not okay.'"

Then an acquaintance in Shanghai helped Feng get a tourist visa to the United States. Daoqun mocked her. "You haven't graduated from primary school. How can you go to the United States?"

In May 2016, Feng arrived in Los Angeles. A friend of a friend, an Uber driver who also worked in construction and whom Feng called *di di,* or "little brother," picked her up. She was hired at a nail salon, a restaurant, and, within a few days, a massage parlor.

As soon as she could, Feng, now going by the name Coco, called home from an American telephone number. No one answered, believing it was a scam. Feng rarely shared details of her life in America with her family back home. Instead, the monthly phone calls centered on the various financial needs of the extended family. Every few weeks, Feng would wire 1,500 yuan (about $230) via WeChat, a free messaging app popular in China, to Daoqun, who would then send the money to their mother in the countryside. Over the years,

Feng paid for her mother's eye surgery; her nephew's school fees; her sister-in-law's business expenses; and the weddings and funerals of relatives and neighbors. Without being asked, Feng always sent extra money for such holidays as Lunar New Year and the Dragon Boat, Mid-Autumn, and Hungry Ghost festivals. She also paid to renovate her parents' house, as well as for the mortgage on her oldest brother's house, which he shared with his wife, his son, and his son's wife. In May 2020, Feng made a down payment on a four-bedroom apartment for her mother. At various times, Feng supported 10 members of her family.

On March 14 last year, around 10 or 11 p.m. ET, Feng called Daoqun to discuss Qingming Jie, an upcoming tomb-sweeping holiday honoring ancestors. Feng would send 1,000 yuan (about $150) so the family could purchase food for the event—two chickens, a goose, some rice, and bananas and apples for dessert—as well as ghost money made from incense paper for burning and customary firecrackers. Daoqun was at the barbershop getting a haircut, so the call was brief.

On March 15 at 10 p.m., Feng called an acquaintance, a government cadre who lived in Feng's mother's village and helped convert the money Feng sent via WeChat into cash. He was in a meeting, and he too rushed Feng off the phone. That would be the last time anyone back home would hear from their little sister.

It wasn't until six days after the killings that Daoqun read what happened on WeChat. He called his sister Mei and asked her to pick up their mother, Huazhen Zhang, and bring her to him in Zhuhai city, near Macao. Daoqun wanted to protect his mother from the news, at least until he knew more. He went to the local police station, where he was given a phone number for the American embassy in Beijing. A staffer confirmed that the anonymous Chinese woman killed at Young's Asian Massage was indeed Daoyou Feng.

Daoqun considered traveling to the United States, but his children dissuaded him from making the perilous and costly journey. The family considered repatriating Feng's body back to China, but according to an ancient local tradition, unmarried daughters who die away from home cannot be buried in their ancestral village.

On April 4, after her body lay unclaimed in the county morgue for 19 days, Feng was at last interred. Her funeral was attended by sympathetic strangers, no friends or colleagues, many of whom were asylum seekers or of otherwise precarious immigration status.

On the same day, Daoqun's wife rose at 4 a.m. to pluck the chickens before the family headed to sweep their ancestral tomb. Arriving at the grave site, they cleaned it of wild growth and lit their ghost money on fire. Others whispered prayers to the long list of spirits who had come before—a list that now included Feng—but Daoqun did not. "I never believed the dead could listen to the living."

THE FIRST MIGRATION

Chinese laborers in the South were among the earliest Asians to migrate to the U.S. from the mid to late 19th century. Anxious white plantation owners hired them during Reconstruction. The first Chinese in Georgia came as contract laborers in 1873, when an Indianapolis construction company brought in 200 Chinese workers to help build the Augusta Canal. Although Chinese labor completed much of the public infrastructure work in Georgia at this time, including railroads and bridges, according to Emory University history professor Chris Suh, the Chinese population was scrubbed from history, subsumed into the Black-white binary of the American South.

These immigrants aided in the "economic transition from raw extraction to something approaching industrial capitalism," as Alexander Saxton writes in *The Indispensable Enemy,* but were reduced to their basic economic function—treated later as "high-tech coolies," says Mount Holyoke College associate professor Iyko Day, pointedly using the slur, derived from the Tamil word kuli, as in "wages." South Asian and Syrian merchants traveled across the American South into the early 20th century, hawking rugs and fabric, or chinoiserie. Then Methodist missionaries began recruiting students from Korea, Japan, and the Philippines to study at Duke, Emory, and Vanderbilt universities.

The Immigration and Nationality Act of 1965, known as Hart-Celler, ended the quota-based immigration system and specifically encouraged immigration from Asia and Africa. The Asian population in America grew from 63,000 people in 1870 to 12 million in 2000, and that number has nearly doubled since. They came to America as members of educated professional classes who, in the new country, became Gujarati hotel operators, Korean shopkeepers, Vietnamese nail salon owners, and Hmong chicken farmers.

III. TAN

Before America, Xiaojie Tan was the second of two daughters born to a bicycle repairman from Nanning, about 170 miles inland from Feng's hometown. Tan's family was similarly poor, and when she was 20, they married off their younger daughter to a shoe salesman. Tan and her husband had a daughter before divorcing. In the early 2000s, Tan met an American roofing businessman named Michael Webb, whom she married in 2004. Two years later they moved to Florida, where Tan worked at a nail salon. In 2010, they moved to Georgia, where Webb was from, and settled in Marietta, where Tan opened her own nail salon. They divorced in 2012, the same year Tan became a U.S. citizen. In 2013, she married Jason Wang, a former foreign student from northern China. In 2016, Tan sold the nail salon and the following year opened Young's Asian Massage 10 miles north of Marietta. In November 2018, Tan and Wang divorced. Two months later, they were back together and planning to marry again.

Wang worried for Tan, who stayed late at the spa most nights. He encouraged her to keep her handgun on her when she went to work. But Tan was afraid of guns and kept it at home, under her pillow.

March 16 had started out as a "nothing special" day, Wang recalls. Tan left for work at 8 a.m. Around 3:40 p.m., she greeted a regular at the spa. Tan ushered him past the room where Lyon and Feng were, into another room on the left side of the hallway.

But this day wouldn't be like the days that had come before. When the man got up, he refused to tip. Tan protested. After he dressed and used the restroom, he began shooting.

IV. YOYO

Chingching, or Yoyo, as her clients called her, figured the first gunshot was not a gun at all but someone heating up a late lunch in the microwave. When she heard the second pop, she opened the door and saw Tan and Feng lying on the floor. Yoyo shut the door and, pressing her slight frame against the plywood, told Mario González, her client, to get dressed. The killer was on the other side, trying to push it open. González joined Yoyo in barricading the door. His wife—who had been getting a massage from Yoyo's colleague Apple in another room—would be among the dead. The gunman left before police arrived.

Yoyo, Apple, and a third spa worker called Jenny also fled,

stopping by their boss Tan's house, where they had been living, before heading for Flushing, New York. With González in handcuffs, there was a sense of an ending at least. No one knew that it was only the beginning, that the violence was far from over, that the killer was still on the loose. After leaving the spa at 4:50 p.m., the killer got on I-75 and headed south, toward Atlanta, toward the Cheshire Bridge spas.

CHESHIRE BRIDGE

Prior to hosting the Summer Olympics in 1996, Atlanta launched a "cleanup" campaign. The city reportedly arrested some 9,000 residents—mostly poor, mostly Black—in the year running up to the ceremonies, mostly on loitering charges. Local officials moved about 6,000 residents out of public housing and gave homeless people one-way bus tickets out of the city. Thousands of residents were relocated to an industrial area in southeast Atlanta known as The Boulevard. Still others were pushed into another industrial area in the city's northeast, near Cheshire Bridge. The area had been no more than a desiccated lot near a rail yard where four train lines met until strip clubs and massage parlors began opening there.

Today the area is crowded with 300-unit apartment buildings, half-million-dollar condos, and the second most popular club in Atlanta, which pulls in $600,000 a week, according to someone familiar with Atlanta's economy. The original inhabitants feel they are being "squeezed out," while newcomers feel besieged by the noise, the traffic, and the specter of illicit activity. Cordoned by I-85 to the north, the upper-middle-class Morningside–Lenox Park neighborhood to the east, a rail junction and storage yard to the west, and Piedmont Park to the south, the area has seen tension building for some time, with nowhere to go.

V. KIM

On March 14, a Sunday, Hyun Jung Kim Grant, 51, prepared a beef bulgogi marinade and set it in the fridge at her home in Duluth, a suburb north of Atlanta that is home to one of the fastest-growing Korean populations in America. She told her sons, Randy and Eric, that she would be home by the end of the week.

Back in the old country, the Kims ran a low-budget guesthouse—a spare room with a shared bathroom—in Gyeongju, a historic and coastal city in the southeastern part of the Korean peninsula. Kim did well in school, and she was sent to the capital, Seoul, for her studies, a chance afforded to only the brightest of students. According to her brother Hyun Soo, Kim attended Dongguk University, a four-year Buddhist research college that has produced many of South Korea's police administrators and K-pop stars.

After college, Kim began teaching middle-school home economics but left to support the new family business: a sushi restaurant in a department store. A salesman working in menswear a few floors below began coming up to take his lunch at the restaurant and noticed Kim, who was "so smart and pretty and easy to talk to and well known by everyone," according to Kim's brother. They were soon married. But following the 1997 IMF crisis—as it's known in Asia; in the West, it's "Asian contagion"—the restaurant closed.

A WORLD, ROCKED

The Asian financial crisis of 1997 began when Asian economies dropped their financial controls to cover deficits caused by export-only growth strategies, resulting in currencies being devalued by as much as 70 percent. The International Monetary Fund stepped in, lending more than $110 billion with strict preconditions such as increasing interest rates, reducing public spending, and other banking sector restructuring, leading to a dramatic collapse in the standard of living for millions across Asia.

—

KIM AND HER HUSBAND DECIDED TO JOIN KIM'S SISTER IN AMERICA. THE PLAN was to travel on a tourist visa and work for a few years to earn U.S. dollars to bring home. The newlyweds left for Washington State in 1998. They settled in Aberdeen, an economically depressed timber and fishing town 100 miles south of Seattle, where Kim's husband picked up shifts at the local laundromat and restaurant, relatively fixed opportunities available to him as a working-class Asian man. Kim went to work at a hibachi, a Japanese grill, where she served the predominantly white clientele in a kimono.

When Kim was pregnant with her first child, she drove past a billboard for an insurance broker named Randy and thought, That's a nice name. Two years later, Kim had another son, Eric. Soon afterward, she and her husband divorced. In 2002, Kim married her aunt's former son-in-law. But by 2008, single again and looking for a second chance, she and her sons headed to Georgia. The family lived out of hotel rooms for a while, until Kim persuaded an acquaintance to take care of her sons, including for a period of a year, when she sought work in other states. She returned, but Randy and Eric seldom saw their mother. Kim was away a few days to a few months, for a total of a third of the year.

In 2014, Kim was getting ready to leave on one of her trips when Randy, then 14, demanded to know what she did for work, really. Kim had told her sons she did makeup, but Randy didn't know any other makeup artists who worked overnight. Kim admitted that she worked at a massage parlor. "I don't know why you thought I would think of you less for it," Randy told her. "It's work. Would you rather be homeless?"

When in town and unwinding from her punishing spa hours, Kim went out at least twice a week, returning as late as six in the morning on weekdays. She would call ahead to Randy, who she knew would be up playing some video game, so he could open the door for her, help her out of her shoes, and tuck her into bed. On such occasions, Kim was in the habit of asking Randy, "Do you know I love you?" Another familiar refrain: "If you are married and have a family, will you let me live with you?" "It was so awkward," Randy told me. "I'm just like, I'm in high school." When Randy became old enough to work, he found a job at a Korean bakery near the local H Mart, a Korean grocery chain that was one of five Asian grocery stores along the Duluth main street, and began pitching in to pay for the gas and internet bills.

That Tuesday, after 5 p.m., Randy was at home on his day off from his bakery job when a text came from the daughter of his mother's coworker Eunja Kang, who went by the spa name Yena. *Did you hear what happened?* Randy pulled away from his computer screen. *Your mom was shot.*

Randy rushed to pick up Eric, who was working as a cashier at a takeout-only Chinese restaurant, and headed for the spa. Eric cried the entire way. At the spa, an officer directed them to the police station, and there, waiting to be interviewed by a homicide detective, they received a call from Kang, who told them their mother was dead.

SPA WORK

The occupation is as common in immigrant communities as it is misunderstood. According to Georgia state human-trafficking awareness training, people with limited English skills living at their place of work is considered a sign of sex trafficking, yet these are standard practices among workers. The work itself might mean ordinary massages, or it might mean massages that include erotic services—specifically manual stimulation, which some workers do not think of as sex work, as it doesn't involve penetration.

Workers like Kim can make as much as $20,000 in a good month. That money supports families in this country and the other. Whatever remains is spent by visiting "room salons" or "host bars," part of a larger world of night culture that originated in Japan and became popular across Asia and in diaspora communities called mizu shobai, the "water trade," where hosts and hostesses lit cigarettes, poured drinks, and provided sexualized company while encouraging patrons to spend more, for which they received a cut. Host bars traditionally catered to men looking for female companions, but in recent times, bars catering to female customers have sprung up, which spa workers often patronized. Money also flows into private gambling dens, where workers get together to play Go-Stop, a Korean card game, or participate in kye, a kind of kitty, meaning "bond," the informal lending system used by many immigrants with no access to official banking systems. Kye has been crucial to newcomers who are locked out of traditional labor markets due to a lack of language skills or discriminatory practices, and wish to start their own businesses. Ivan Light of UCLA estimated that as much as 40 percent of Korean-owned businesses in Los Angeles have been financed via kye, which has a social as much as an economic function, and together with the water trade is among the few ways people like Kim had of staving off the incurable loneliness that is central to immigrant life.

VI. PARK

Despite her 12-hour workday (14 including the commute), Soon Chung Park preferred spa work—greeting guests, doing laundry, preparing meals and snacks for her colleagues at Gold Spa—to nearly any other work she'd undertaken since immigrating to the United States from South Korea in 1986, including jobs at a deli, a restaurant, and a farm. She had also dealt in diamonds, giving her the nickname Jewelry Park, which had settled into the English name Julie, as she was known by most everyone in her new life.

Back in Korea, after her family finances collapsed, she "ran away" to New Jersey, where her older sister lived. Over the years, Park's five grown children joined her, opening a sushi restaurant, a bodega, and a nail salon in New Jersey and in New York, where they still lived. After declaring bankruptcy in New York in 2013, Park fled from the lips of ruin to Georgia, where she became a frequent visitor to host bars. In 2017, she met Gwangho Lee, a 38-year-old bar host who had come to the U.S. in May 2015 on a three-month tourist visa.

In June 2018, Park and Lee were married. The next year, Lee submitted his paperwork for a green card. The owner of a spa served as a sponsor. Because he was 36 years younger than Park, Lee was referred to as "boy groom."

In August 2018, Park was arrested during a vice raid at a spa in an Atlanta suburb and charged with two counts of keeping a place of prostitution, both of which were dismissed, and convicted of one count of criminal trespass. She spent a month under house arrest wearing an ankle monitor, which she herself paid for. Park told Lee that another worker had been turning tricks, and that she got caught up in the raid. Later that year, she began working at another spa, called Gold, which had its own run-ins with the law, according to *The New York Times.* It had been the site of seven stings from 2011 to 2014, where at least 11 workers were arrested for prostitution-related charges, including keeping a place of prostitution, masturbation for hire, and offering sex acts to undercover cops for as much as $400.

Soon after Park started at Gold, Lee also began working there, running errands for the staff when he wasn't driving a taxi or painting houses.

On March 16, Lee was tasked with giving another spa worker, Eunja Kang, a ride home. On his way to Gold, before 5 p.m., Park called, wanting to know what was taking so long. Anyway, she had to go, a customer had arrived, she said. As Lee neared the spa at 5:45 p.m., Kang sent a string of texts, telling Lee the spa had been "robbed" and that his wife had "fainted." Arriving at the spa, Lee saw his wife lying on the floor, her dentures chipped from the fall.

Minutes later, Lee pieced together what happened. The killer had first shot Suncha Kim, 69. Park must have stepped out of the kitchen and was shot next, followed by Hyun Jung Kim Grant. Eunji Lee, 41, was napping in a nearby room. Kang, 48, was in the main room waiting for Gwangho Lee, her ride home.

When the shooting began, Kang opened the door. The killer was standing ahead, staring back at her. She shut the door and took cover under the duvet. Eunji Lee hid behind a large box in the same room. The door opened. Kang heard two shots, but they missed the women.

LEAVING

Before the immigrant becomes an immigrant, before this single act comes to define her, she is preoccupied with what lies ahead. She knows that this leaving will take her away from home. But what she often does not know is that folded into the decision to go away is also the decision to potentially never see her family or homeland again. On one side of the scale, she has put the sum of her life thus far. On the other is America and some vague yet hopeful feeling that life will be better there. And because she has to, or because she wants to, she chooses that one vague and hopeful feeling over everything else—an act that speaks to the vast and violent inequalities that exist in the world.

VII. YUE

The killer left Gold and crossed the street, entering Aromatherapy Spa, where Yong Ae Yue, 63, was working. Yue had met her husband, Mac Peterson, an American G.I., in 1976 while selling commuter-train tickets between Seoul and the southern port city of Busan. The couple had a son, Elliott, in 1978, and later that year, when Peterson was reassigned to Fort Benning, Georgia, the family moved there. In 1982, their second son, Robert, or Bobby, was born, but by 1984, Yue and Peterson had divorced. The boys moved with their mother to Galveston, Texas. In 1987, Yue transferred custody of her two sons to Peterson, which is how the boys came to live with their father in Georgia. After a decade away, Yue joined her family in Georgia, where she picked up odd jobs, mostly at spas. In 2008, Yue was charged with two prostitution-related offenses. Yue, like Park, told her family that another worker at the spa had been involved in prostitution and she had gotten caught up in the raid.

In late 2020, Yue began working at Aromatherapy Spa, where one of her responsibilities was greeting guests. And so, at around

6 p.m. on March 16, when the bell rang, Yue was ready by the door, opening it, hello. She was shot in the face.

CHEROKEE COUNTRY

Woodstock, Georgia, was Cherokee country before its original inhabitants, who had been in the area for 11,000 years, were displaced by white settlers around the mid-1700s. On May 28, 1830, President Andrew Jackson signed the Indian Removal Act, codifying into law the forcible removal of 15,000 Cherokee people from what is today their namesake county. The white settlers panned for gold in nearby rivers, purchased Black people as slaves, and opened chicken-processing plants, still in operation nearly two centuries later.

Woodstock today enjoys a median family income of $76,191, and is almost 80 percent white. It is the hometown of at least two notable figures: Dean Rusk and Eugene Booth. Rusk, who later became secretary of state, was responsible for splitting the Korean peninsula in two using a foldout map from a copy of *National Geographic*. The line "made no sense economically or geographically," he later admitted, but it allowed American occupying forces to take control of Seoul, a decision that would divide families for generations. Booth was a nuclear physicist and core member of the Manhattan Project, which led to the bombings of Hiroshima and Nagasaki in Japan, killing as many as 250,000 civilians, according to some estimates. Woodstock is proud of their native sons, naming a middle school after Rusk.

VIII. THE SUSPECT

The distance between where the workers lived and where the suspect is from is no more than 20 miles, but no major highways connect the two communities. One must drive on undulating country roads to get from one to the other. The suspect was born on April 6, 1999, to a father who ran a lawn care business in the area and a mother involved at the Crabapple First Baptist Church, which the family attended every Sunday. He grew up in a house in Woodstock, 30 miles north of Atlanta; attended high school five miles north in Canton, which gets its name from the Portuguese word for the Chinese port city of Guangzhou: Cantão—Canton in English; and attended church in Milton, five miles south. Not long before March 16, the suspect's parents had kicked him out for watching porn. He moved in with a friend from church. On the morning of March 16, he stayed home from his landscaping job—inclement weather—and watched porn. The church friend had confronted him, and the suspect had left the house, ashamed.

When the suspect arrived at Big Woods Goods, an employee ran an instant background check. The suspect had no criminal record. Minutes later, he walked out with a 9-mm handgun.

He was a typical mass shooter in that he was white and male. He was unusual in his age—21; the average is 33—and in the fact that, unlike 60 percent of American mass shooters, he did not appear to have a violent history, nor any prior convictions, at least none in the public record. There had been no known childhood trauma, either.

He was a product of his social world, including the Cherokee County School District, which he attended from kindergarten to 5th grade, and again from 7th to 12th grade, graduating from Sequoyah High School in 2017, though nobody recalled him with any kind of ringing clarity or enthusiasm.

One former Sequoyah student, Sydney Rosant, class of '19, says her school's defining trait is its "culture of intolerance." Each year, a committee of teachers chooses a senior student from a pool of applicants to be "chief" of the school spirit squad; that student was required to dress as a Native American chief. As of the 2020–21 school year, the costume is no longer a requirement. Trey Brown, class of '19, who now works at Chili's Grill & Bar in Brookhaven, recalls that the Confederate flag was everywhere: on backpacks, belt buckles, car decals. In 2020, the school district sent out a statement addressed to students to not display Confederate flags at school and this also applied to the dress code. Rosant and Brown, who are Black, do not recall being taught how the school got its name, nor the area's history of violence. All they were taught about race in America, Brown recalled, was that "MLK did this and this and this and now racism is gone." (A spokesperson for the school district points out that Georgia's educational curriculum requires lessons on Sequoyah, the Trail of Tears, and racism.)

THE POLICE

One Saturday afternoon in June, I drove over to the suspect's parents' house, a late '70s single-story slate-gray ranch house surrounded by maple, red oak, and white pine, at the beginning of a dead-end road. It had been three months since the shootings, before the suspect would plead guilty to the first four murders and before he would plead not guilty for the other four, charges for which he still awaits trial. The plea deal lets the suspect avoid the death penalty, which continues to be pursued for the remaining charges. The DA has called the killings hate-related crimes—acts that further exacerbated well-founded fear among Asian Americans. The suspect has told authorities he was motivated by, in his words, sex addiction.

I rang the bell at the family's home. No one answered. Before I could decide what to do, a police cruiser showed up. An officer who introduced himself as Sergeant Clement explained that the neighbors—multiple people—had called to report "suspicious activity."

"The one good thing about Cherokee County," he told me, "is that we look out for each other. It's like how it used to be in the '70s."

I asked Clement what, specifically, the neighbors were worried about. "To be honest," he said, "what they are worried about is . . . they are afraid of revenge."

IX. JEONG

Reporting in that community, I was seen as an emissary from the other world. But of course, I did not belong to the other world, either. Beyond certain biological features, I had nothing in common with the women I was writing about. And yet, as on one side of the river I had been met with inexplicable animus, on the other side was an inverse sense of proximal affinity. People I spoke with in the Korean and Chinese communities at times did not see me as a reporter. Some of them called me by my Korean name, pronouncing every syllable correctly—and this unexpected grace thrilled me. They asked me to translate for them, give them a job, pick up a dog from the airport and foster it for the weekend. I said no to all but one.

When the plague of 2020 came and I was no longer heading out on reporting trips, I had begun directing my questions to two subjects who had eluded me the most: my parents. Among the first questions I posed was one about a table we had in our hallway when I was growing up. An odder of my childhood afflictions was a dream of becoming Helen Keller. I would walk around the house pretending

I was deaf and blind, forcing my parents to begrudgingly re-childproof their postmodern furniture, including the hall table in question, with its sharp corners. After bringing up the subject of the table—I only wanted to know what had happened to it—I noticed that our family WhatsApp group had gone silent.

When I confronted my father, he told me he had given it away after his business closed as a result of the 1997 crisis (something I hadn't known), and that my bringing it up had upset him and my mother so much that he had been banished to the guest bedroom. There were other secrets that I will spare my family from disclosing here. He was crying as he explained that this—the pain he was so plainly demonstrating—was why he wanted to opt out of whatever it was I thought I was doing. Refusing to be narrativized, in the language of the empire no less, was an act of resistance for him.

I felt I was locked in an ideological battle—my advocating for excavation, my father choosing obfuscation. But now, watching my father's pixelated crying on my iPhone, I was filled with doubt. "Only asking questions," and writing, are violent acts. When doctors inflict pain, it is with a promise of a cure. I wasn't sure what I was offering in exchange for dredging the past for its harms.

X. CHUNJA

In the years between the end of the Japanese occupation and the beginning of the civil war in the Korean peninsula, ethnic Koreans in Japan, called Zainichi Koreans, many of whom had been taken as slaves, returned to their newly liberated homeland. Fifteen-year-old Chunja was among them.

Soon she found work at a shipping company in Busan, where she met a young customs official named Suk Myung, the youngest son of a prosperous land-owning family from the north. His mother had sewn gold bars into the lining of his coat, and Suk Myung had set off on foot, heading south. Not long after, Dean Rusk accessed his *National Geographic,* stranding Suk Myung in the south with a northern accent that marked him as an outsider, and severing him from his family.

Chunja and Suk Myung married, and soon Chunja was pregnant. Before Chunja could give birth, however, Suk Myung was sent to sea on a mission to thwart pirates and was caught in a storm. Chunja received a handwritten letter informing her that her husband had been killed. The letter, which she would keep until she died, included the rough coordinates of where the ship had capsized. Chunja, a widow and soon a single mother to a baby boy, began supporting her family by selling laundry soap. She did well for a while, teaching herself English and setting up an import-export business, which made her enough to pay for such markers of privilege as chocolate bars and powdered milk—sold by peddlers hawking goods that had fallen off the back of trucks from the nearby U.S. military base, one of 100 across Asia-Pacific.

When Chunja's son grew up—by now the family was living in Daegu, where Chunja would settle—and announced that he would like to go to art school, she sent him off to study in America.

But her son had met and married a woman from his English class, and she had gotten pregnant with their first child. After the woman's application for an American visa to join her husband was denied, she tried again and was denied a second time. Then she gave birth. On the third try, she was advised to leave the infant behind, a kind of collateral for her return. This time the visa came through, and she left for America, leaving her newborn in the care of her maternal grandmother.

I am that child.

XI. RANDY AND ERIC

On May 4, 49 days after the shootings, Randy and Eric went to visit their mother. Buddhists believe that on the 49th day after a person dies, their spirit moves on to one of three realms: the hell realm, a heavenly kind of realm, or this realm. According to *The Tibetan Book of the Dead,* or the *Bardo Thödol,* whence this tradition comes, the soul must be helped along as it passes into the next world. May 4 also happened to be Kim's birthday as well as Eric's, so they had Korean fresh cream cake.

Between a nominally nicer, more expensive all-white cemetery and a cheaper and more diverse option, the brothers chose to bury their mother along with other people of color. "I don't think she would be comfortable with all white people," Randy told me. Even in death, America remains segregated.

Death was costly. For as long as anyone could remember, Kim had been boasting that she had a big kye payout coming, about $100,000, in the fall. When she died, however, the balance in her checking account was $200. So Randy was grateful for the donations—nearly $3 million—that had come after the killings.

Back in China, Daoqun had continued to lie to his mother, telling her America was having "Wi-Fi problems," which was why Feng hadn't called home in many months.

Hernandez-Ortiz, the mechanic who was shot in the face, had the great fortune of surviving but the misfortune of doing so in

America, where his health insurance did not cover the multiple surgeries he has had to undergo. In the first three months of what was to be a long recovery, he'd incurred half a million dollars in medical debt. Before everything, Hernandez-Ortiz had been an avid singer, performing Guatemalan folk songs. Now he drinks from a straw and speaks in a scratchy, raspy voice.

So much of our fate is luck. "We don't choose our parents," says Chris Suh, the Emory history professor. "We were born wherever we were born, and that often determines how our future gets made."

His comment conjures the "veil of ignorance," the philosophical device by which being blind to one's own status negates bias. Philosopher John Rawls believed that under this "original condition," as the veil is also known, everyone would aim for as fair a society as possible. If you were building this platonic ideal world and you didn't know if you would end up a spa worker, or a FedEx driver, or an undocumented worker, or a Black student in a predominantly white school in the Deep South, or a Cherokee tribesperson in 1830, or a Chinese laborer in xenophobic gold-rush California, what world would you want to build?

HOME

In the triangle where Gold and Aromatherapy spas had been, there was a third spa called ST Jame. On a predictably blazing summer afternoon, a worker from ST Jame was taking a break outside. She was 77 years old and had come to the United States from South Korea after marrying an American soldier some 40 years ago. They never learned to get along, she told me, and after the divorce, she began working at parlors to support herself. She made $100 a day, she said. She cooked, cleaned, did the wash, and opened the door for patrons. The other women who worked there rented rooms and saw clients for $80 per two-hour session. Business had slowed since March. I asked her if she wasn't afraid. Yue at Aromatherapy had died opening the door for a customer, after all.

"Why scared?" she asked. "People can die anywhere." When it came time to shuffle off this mortal coil, she told me, she planned to go back to South Korea. It would be like your soul coming back to your body, she said. It would be like bringing your body back home, like coming home to your own body.

ORIGINALLY PUBLISHED IN *VANITY FAIR*, MARCH 2022

PART II

THE TRUE CRIME STORIES WE TELL

WHO OWNS AMANDA KNOX?

BY AMANDA KNOX

Does my name belong to me? Does my face? What about my life? My story? Why is my name used to refer to events I had no hand in? I return to these questions again and again because others continue to profit off my identity, and my trauma, without my consent. Most recently, there is the film *Stillwater,* directed by Tom McCarthy and starring Matt Damon and Abigail Breslin, which was, in McCarthy's words, "directly inspired by the Amanda Knox saga." How did we get here?

In the fall of 2007, a British student named Meredith Kercher was studying abroad in Perugia, Italy. She moved into a little cottage with three roommates—two Italian law interns, and an American girl. Less than two months into her stay, a young man named Rudy Guede, an immigrant from the Ivory Coast, broke into the apartment and found Meredith alone. Guede had a history of breaking and entering. A week prior, he had been arrested in Milan while burglarizing a nursery school, and was found carrying a 16-inch knife. He was released. A week later, he raped Meredith and stabbed her in the throat, killing her. In the process, he left his DNA in Meredith's body and throughout the crime scene. He left his fingerprints and footprints in her blood. He fled to Germany immediately afterward, and later admitted to being at the scene.

I am the American girl in that story, and if the Italian authorities had been more competent, I would have been nothing more than a footnote in a tragic story. But as in many wrongful convictions, the authorities formed a theory before the forensic evidence came in, and when that evidence indicated a sole perpetrator, Guede, ego and reputation led them to contort their theory to maintain that I was still somehow involved. Guede was quietly convicted for participating in the murder in a separate fast-track trial, and then I became the main event for eight long years.

While I was on trial for the murder of Meredith Kercher, from 2007 to 2015, the prosecution and the media crafted a story, and a doppelgänger version of me, onto which people could affix all their uncertainties, fears, and moral judgments. People liked that story: the psychotic man-eater, the dirty ice queen, Foxy Knoxy. A jury convicted my doppelgänger, and sentenced her to 26 years in prison. But the guards couldn't handcuff that invented person. They couldn't escort that fiction into a cell. That was me, the real me, who returned to that windowless prison van, to those high cement walls topped with barbed wire, to those cold, echoing hallways and barred windows, to that all-consuming loneliness.

—

TEN YEARS AGO, AT THE AGE OF 24, I WAS ACQUITTED, AND I TUMBLED INTO A kind of purgatory. I left one cell and immediately entered another: the quiet of my childhood bedroom. Outside, the telephoto lenses were fixed on my closed blinds. Prison had given me an appreciation for all the freedoms I'd taken for granted. Freedom showed me how many I still lacked.

As I walked back into the free world, I knew that my doppelgänger was there alongside me. I knew that everyone I would ever meet from then on would have already met, and judged, her. I had been acquitted in a court of law, but sentenced to life by the court of public opinion as, if not a killer, then at least a slut, or a nutcase, or a tabloid celebrity. *Why doesn't she just go away already? Her 15 minutes are over.*

In freedom, I had become a pariah. Looking for work, going back to school, buying tampons at the pharmacy, everywhere I went I met people who already thought they knew who I was, what I'd done or not done, and what I deserved. I was threatened with abduction and torture in broad daylight; I was threatened with having Meredith's name carved into my body. Strangers sent me lingerie and bizarre love letters. All over the world, people believed they knew me, a warped assumption that turned me into a monster to some and a saint to others. I felt like I was always standing behind that cardboard cutout, Foxy Knoxy, saying, *Hey, back here, the real me!* Even most of the strangers who offered kindness and support didn't truly see me. They loved her.

It's hard to make friends, to date, to be a regular person when everyone you meet has a preconceived notion of who you really are, whether positive or negative. I could have chosen to hide out, to change my name, to dye my hair, and hope no one recognized me ever again. Instead, I decided to embrace the world that had dehumanized me, and all those who turned me into a product.

From the moment I was arrested, my name and face and trauma became a source of profit for news organizations, filmmakers, and other artists, scrupulous and unscrupulous. The most intimate details of my life, from my sexual history to my thoughts of death and suicide in prison, were taken from my private diary and leaked to journalists. Those journalists turned my darkest fears into fodder for hundreds of articles, thousands of blog posts, and millions of hot takes. People speculated about my mental state and sexuality, they diagnosed me from afar, they used my predicament as a metaphor, they made TV movies about me, based characters in legal shows on me, and the worst of them took every opportunity they could, while I was in prison and while I've been out, to shame me for something I didn't do, to shame me for living while Meredith is dead, to shame me for being in the very headlines they write, for being in the photographs they take without my consent. The hypocrisy and the cruelty are maddening. And yet, being under that microscope has given me insight into how wrong a media narrative can be, how easy it is for all of us to consume other people's lives as if they were mere content to fill up our Twitter feeds.

All of this has led me to dedicate myself, in my written journalism and on my podcast, *Labyrinths,* to upholding the ethical principles I so often found lacking in the media that covered and consumed me. I believe that journalists must always center people in their own stories, and recognize what is at stake for their subjects. But even as I put my own voice into the world, the *idea* of me is still an object for others to consume. So I can't say I was surprised when I heard about Tom McCarthy's new film.

Stillwater is both "loosely based on" and "directly inspired by" the "Amanda Knox saga," as *Vanity Fair* put it in an article published by a for-profit magazine company promoting a for-profit film, neither of which I am affiliated with. I want to pause on that phrase, "the Amanda Knox saga," because the manifold ways that my identity continues to be exploited start with this shorthand. What does "the Amanda Knox saga" refer to? Does it refer to anything I did? No. It refers to the events that resulted from the murder of Meredith Kercher by Rudy Guede. It refers to the shoddy police work, the flawed forensics, and the confirmation bias and tunnel vision of the Italian authorities whose refusal to admit their mistakes led them to wrongfully convict me, twice. In those four years of wrongful imprisonment and eight years of trial, I had near-zero agency.

Everyone else in that "saga" had more influence over the course of events than I did. The erroneous focus on me by the police led to an erroneous focus on me by the press, which shaped how I was presented to the world, and continues to shape how people treat me today. In prison, I had no control over my public identity, no voice in my own story.

This focus on me led many to complain that Meredith Kercher had been forgotten. But whom did they blame for that? Not the Italian authorities. Not the press. Somehow it was my fault that the police and media focused on me at Meredith's expense. The result of this is that 14 years later, my name is the name associated with this tragic series of events I had no control over. Meredith's name is often left out, as is Rudy Guede's. When he was released from prison in late 2020, the *New York Post* headline read: "Man Who Killed Amanda

Knox's Roommate Freed on Community Service." My name is the only name that shouldn't be in that headline.

—

IN LIGHT OF THE #METOO MOVEMENT, MORE PEOPLE ARE COMING TO UNDER-stand how power dynamics shape a story. Who had the power in the relationship between Bill Clinton and Monica Lewinsky, the president or the intern? Shorthand matters. Calling that event "the Lewinsky scandal" fails to acknowledge the vast power differential, and I'm glad that more people are now referring to it as "the Clinton affair," which explicitly calls out the person with the most agency in that series of events. I would love nothing more than for people to refer to the events in Perugia as "the murder of Meredith Kercher by Rudy Guede," which would make me the peripheral figure I always was, the innocent roommate.

But I know that my wrongful convictions, and my trials, became the story that people obsessed over. I know they're going to call it "the Amanda Knox saga" in perpetuity. I can't change that, but I can ask that when people refer to these events, they make an effort to understand that how you talk about a crime affects the people involved: Meredith's family, my family, my co-defendant, Raffaele Sollecito, and me.

Every review of *Stillwater* I have seen has mentioned me, for better or worse. Some refer to me as a person convicted of murder, while conveniently leaving out the fact of my definitive acquittal. *The New York Times,* in profiling Matt Damon, referred to these events as "the sordid Amanda Knox saga." That's not a great adjective to have placed next to your name, especially when it describes events that you didn't cause, but that you suffered from.

Even the tiniest choices people make about how to refer to newsworthy events shape how they are perceived. And when real events serve as inspiration for fiction, that effect can be magnified and distorted. *Stillwater* is by no means the first project to use my story without my consent, and at the expense of my reputation. While I was still in prison, and still on trial, Lifetime produced a film called *Murder on Trial in Italy.* I sued the network, which resulted in it cutting from the film a dream sequence that depicted me killing Meredith. A few years ago, there was the Fox series *Proven Innocent,* starring Kelsey Grammer, which was developed and described as "What if Amanda Knox became a lawyer?" The first time I heard from the show's makers was when they had the audacity to ask me to help them promote it on the eve of its premiere.

My name, my face, my story have effectively entered the public imagination. I am legally considered a public figure, and that leaves me little recourse to combat depictions of me that are harmful and untrue, and gives carte blanche to anyone who wants to write about me to do so without consulting me in any way. A prime example is Malcolm Gladwell's book *Talking to Strangers,* which features a whole chapter analyzing my case. He reached out to me just before publication to ask if he could use excerpts of my audiobook in his audiobook. He didn't think to ask for an interview before forming his conclusions about me. I allowed him to use my voice, because at least Gladwell was arguing for my innocence. But even he put the burden of my wrongful conviction on me, based on my behavior, rather than on the authorities, who held all the power in that dynamic. To his credit, Gladwell responded to my critiques over email, and was gracious enough to join me on my podcast. I have extended the same invitation to Tom McCarthy and Matt Damon.

McCarthy was "inspired by" my story, and told *Vanity Fair* that "he couldn't help but imagine how it would feel to be in Knox's shoes." But that didn't inspire him to ask me how it felt to be in my shoes. He became interested in the family dynamics: "Who are the people that are visiting [her], and what are those relationships? Like, what's the story *around* the story?" My family and I have a lot to say about that, and would have happily told McCarthy if he'd ever bothered to ask. McCarthy has no legal obligation to do so. And he is, after all, telling a fictional tale. But legal mandates are not the same as moral or ethical ones.

"We decided, 'Hey, let's leave the Amanda Knox case behind,'" McCarthy told *Vanity Fair*. "But let me take this piece of the story—an American woman studying abroad involved in some kind of sensational crime and she ends up in jail—and fictionalize everything around it." But that story, my story, is not about an American woman studying abroad "involved in some kind of sensational crime." It's about an American woman not involved in a sensational crime, and yet wrongfully convicted of committing that crime. It's an important distinction. Now, if he truly were leaving the Amanda Knox case behind, that shouldn't matter. But then why does every review of *Stillwater* mention me? Why do photos of my face, not owned by me, appear in articles about the film?

Clearly, McCarthy has not left the case behind. Which means that how he has chosen to fictionalize my story affects my reputation. I was accused of being involved in a death orgy, a sex game gone wrong, when I was nothing but platonic friends with Meredith. But the fictionalized version of me in *Stillwater* does have a sexual relationship with her murdered roommate. In the film, the character based on me gives a tip to her father to help find the man who really killed her friend. Matt Damon tracks him down. This fictionalizing erases the corruption and ineptitude of the authorities. Truth is stranger than fiction here: In reality, the authorities already had the killer in custody. Rudy Guede was convicted before my trial even began. They didn't need to find him. And even so, they pressed on in prosecuting me, because they didn't want to admit they had been wrong.

Vanity Fair reported that "*Stillwater*'s ending was inspired not by the outcome of Knox's case, but by the demands of the script [McCarthy] and his collaborators had created." I read a sentence like that, and I immediately wonder: Is the character based on me actually innocent in McCarthy's film?

It turns out, she asked the killer to help her get rid of her roommate. But she didn't mean it *that* way. She wanted her roommate evicted, not killed. Her request indirectly led to a murder. Did McCarthy consider how that creative choice affects me? I continue to be accused of knowing something I'm not revealing, of having been involved somehow, even if I didn't plunge the knife.

By fictionalizing away my innocence, by erasing the role of the authorities in my wrongful conviction, McCarthy reinforces an image of me as guilty. And with Damon's star power, both are sure to profit handsomely off this imagined version of "the Amanda Knox saga" that will leave plenty of viewers wondering, *Maybe the real-life Amanda was involved somehow,* and googling whether the film's story is true.

I never asked to become a public person. The Italian authorities and global media made that choice for me. And when I was acquitted and freed, the media and the public wouldn't allow me to become a private citizen again. I have not been allowed to return to the relative anonymity I had before Perugia. I have no choice but to accept the fact that I live in a world where my life, and my reputation, are freely available for distortion by a voracious content mill.

I'm not angry with Matt Damon or Tom McCarthy, but I am disappointed that they didn't seem to appreciate the value of my perspective and voice, only the value of my circumstances as inspiration and my name as a marketing tool. For four years, I lived alongside women actually guilty of crimes ranging from petty theft to filicide. And let me tell you, playing cards with a drug dealer and being taught to roll out pizza dough with a broomstick by a mafiosa certainly puts things in a new perspective—one that doesn't excuse people's crimes, but puts them into context. I came to recognize the humanity in my fellow inmates, imperfect people whom society had written off as worse than worthless, or as monsters. Those same judgments were, and still are, hurled at me, despite my innocence.

All of this has made me extremely skeptical of those who easily pass judgment. It has made me allergic to the impulse to flatten others into cardboard, to erase their human complexity, to rage against things about which I know only a snippet. Judgment only gets in the way of understanding. Refraining from judgment has become a way of life for me. Call it radical empathy, or extreme benefit of the doubt. I know how wrong people were about me, and I don't ever want to be that wrong about another person. The world is not filled with monsters and heroes; it's filled with people, and people are extraordinarily complex. That includes Tom McCarthy and Matt Damon. I'm sure they had no ill intent. That, too, matters.

ORIGINALLY PUBLISHED IN THE *ATLANTIC*, JULY 2021

TIE A TOURNIQUET ON YOUR HEART: REVISITING EDNA BUCHANAN, AMERICA'S GREATEST POLICE REPORTER

BY DIANA MOSKOVITZ

To hear Edna Buchanan tell it, the phone rang at her desk in Miami on Friday, Dec. 21, 1979, and from that point on, "Life would never be the same again." The call was a tipster reporting that a Black motorcyclist was either dead or about to die as the result of a beating he'd received at the hands of a group of all-white police officers, who'd pummeled him with their long, metal flashlights that doubled as batons. The motorcyclist's name was Arthur Lee McDuffie. If you are of a certain age and grew up in South Florida, you know McDuffie's name and the broad outlines of what followed: McDuffie died, and the officers who killed him were subsequently acquitted at trial by an all-white jury in Tampa, sparking a three-day-long uprising. The way broader America remembers Rodney King, South Floridians remember McDuffie.

But this is not the story of Arthur Lee McDuffie. It's the story of Edna Buchanan.

Buchanan's 1987 book, *The Corpse Had a Familiar Face,* came out at the peak moment of Miami's murder rate, the power of newspapers, and the author's own fame as the country's greatest police reporter. The book is in part a memoir of her life on the police beat for the *Miami Herald,* part victory lap, and part manifesto on what crime reporting should look like. Calvin Trillin had profiled her for the lofty pages of *The New Yorker* the year before, in a piece that itself became a minor legend, making Buchanan both a famous journalist and the subject of a famous piece of journalism. Months later, she won the Pulitzer Prize in general news reporting for her "versatile and consistently excellent police beat reporting." If you believe that the hype around a book can be measured by its review in the *New York Times*—and who writes that review—then know that Buchanan's book was reviewed in the *Times* by the modern master of vampire novels, Anne Rice. She loved it.

Book buyers agreed. A year after being published, the book required a fifth printing. Disney bought the movie rights. The TV movie came out in 1994, starring *Bewitched*'s Elizabeth Montgomery as Buchanan. All this put her in a rarified air few writers, let alone journalists, ever reach.

My copy of *The Corpse Had a Familiar Face* is a first edition, with the bright orange background, fuchsia palm trees, and fake bullet holes that seem to have been required of anything representing Miami in the '80s. I reached for it as America erupted this month, yet again, in protests over the killings of Black people at the hands of police, wondering what Edna Buchanan, one of the greatest influences on late-20th-century crime writing, would have to offer this moment.

But what struck me, from page one onward, was how police positive it was. How it is littered with calls for tougher justice, using victims as props to demand harsher sentences, and how it ignored all the ways American society sets people up to break the law in the first place. How bad behavior by officers—even the one Buchanan briefly married—is condemned, but never really traced back to any larger issue. How Buchanan's words have reinforced institutions that a growing American conscience believes are no longer, and perhaps never were, inherently good, or even necessary at all.

Maybe I could write all this off as a book that just didn't age well, if it weren't for what Buchanan and her body of work have represented and, for decades, set the template for—the epitome of what a crime reporter in America is expected to be.

The book begins with a simple refrain. Everyone moves on from murder, Buchanan declares, except for her. Detectives get assigned other cases. Editors get excited about a new story. The public memory fades . . . but not Buchanan's. Like an oracle predicting a future filled with murder recap podcasts and an endless pipeline of true crime shows to binge, she lays out the appeal of true crime in one sentence: "But I can't forget."

Except memory is a fickle thing, and it's telling what Buchanan remembers from all those years of writing about the dead. There are moments when she seems to be upholding the values of each and every life, like when she says of families that never get justice, the loss of good people too soon: "There is no dirt-bag murder." But elsewhere she offers the startling observation that most people have nothing to fear in Miami, because the vast majority of victims "contribute to their own demise."

"They deal drugs, steal, rob, or stray with somebody else's mate until a stop is put to them," she writes. "Most Miami murder victims have arrest records, most have drugs, alcohol, or both aboard when somebody sinks their ship."

And it's not just the not-so-innocent victims who are at fault. Buchanan deplores people who won't talk to police, who go "blind when the shooting started," without acknowledging the many good reasons why some communities don't trust the police. Some kids will never develop a conscience, she says, detailing the lives of murderous teens, giving no space to the idea that murderous teens are a product of their society. She admits to never forgiving one woman's killer, which might seem noble until you consider that she is neither judge, nor jury, nor the woman's family or friend. She recalls a man found dead, a "nickel-and-dime dope dealer," and mentions in a passing glance how police recognize him—turns out they shot him once. Buchanan doesn't spend a moment wondering if that should be a reason why a police officer, or anyone, should recognize anybody. In her world, bad things happen, and Buchanan does her best to describe the bad things, but she never asks why.

And that's Chapter One.

—

THE HEART OF THE BOOK IS BUCHANAN BREAKING DOWN THE LIFE OF A POLICE reporter. The chapters have self-explanatory titles like "Crooks," "Sex," and "Missing." "Crooks" opens with the advice, "You should not pity most criminals, either; tie a tourniquet on your heart. Sad and sleazy losers are easy to feel sorry for, until you recall what they have done, over and over and over, and will continue to do, given the chance. They say all they need is a break, but if you check it out, you find they've used up lots of them."

"Sex" is mostly about rape, and includes accounts of five officers who sexually assault a teenage sex worker and an ex-cop who murders his sweetheart, although Buchanan doesn't stop to ask if that should have been of concern. "Drugs" is exactly what you'd expect of 1980s writing about drugs—they're bad, don't do them, and they ruined Miami. ("Drugs" also contains the only passing mention of AIDS in the book, in a passage about crack houses.) "Justice" is about what it takes to get bad guys into jail. "Missing" is a template for every pretty-white-girl-goes-missing story you've ever read. Most of it follows mother Susan Billig in her yearslong quest to find her missing daughter, Amy, and it entails private investigators, chasing down tips, and befriending murderous bikers; Susan Billig is the archetypical mother on a mission. But you know the ending of this story because it has, thanks to technology, become an everyday part of the news cycle. It's Nancy Grace, right on the page. Amy's never been found. Her story lives on at her own Unsolved Mysteries Wiki. Susan Billig died in 2005.

Naturally, there's a chapter called "Cops." It's here that the true tension of being a police reporter unfurls. Because to be a great police reporter—the kind editors champion, the kind that gets raises and promotions, the kind that wins a Pulitzer—you have to be friends with a lot of cops.

This sort of capture is true, to some extent, of almost every job in American journalism. To get promotions you need scoops, to get scoops you need sources, and to have sources you need to cultivate a certain camaraderie with people; they have to trust you, which might or might not translate to their belief that you are willing to tell the story their way, and not ask too many uncomfortable questions. But on the police beat, this tension is ratcheted way up. Here, police have already set the narrative. They are where the story starts, they have the official titles and the institutional authority, and also they have the guns.

But the Buchanan model is not, primarily, about police accountability. It's about writing a story that leaps off the page with stunning details, a story that will start with a witty opening sentence—the "lede"—and a career full of brilliant ledes, so many that the debate over which lede was your best will appear in *The New Yorker*. In this model, the goal is to write sentences like "Gary Robinson died hungry," or "Bad things happen to the husbands of Widow Elkin." Something powerful, distinctive, and confident. Something that screams 1A, the analog equivalent of going viral.

Getting that level of detail that fast on a daily basis requires a lot of help. Help getting to a crime scene before anyone else. Help knowing key facts, like what show was playing on the TV at the time of the homicide. Or help closing certain loops in a narrative—what happened in a gap of time, or who fired the gun. There is only one group of people who can consistently provide that level of detail about a crime scene. The Buchanan model requires knowing, and gaining the trust of, a lot of cops.

Buchanan is brutally honest in this regard. She names her favorite cops who are also her favorite longtime sources. Miami police, she writes, "must be among the best and most experienced in the world." She pities the officers who can no longer retaliate when someone spits or curses at them. Florida's long history of police brutality, including influence by the Ku Klux Klan itself, and Miami's specific and not-even-100-years-ago relationship with the Klan, goes unmentioned; all her readers get is the occasional reference to projects at the *Herald* on police brutality. In her account, the bad cops only slipped in due to bad hiring practices, or they gave in to various temptations. And she never dwells on police mistakes. Ever.

Cops are only human, Buchanan reminds her readers, which is true. So is every person she writes about.

—

SHERWOOD "BUCK" GRISCOM WAS "SOUTH FLORIDA'S MOST DEADLY COP," BUchanan informs us. He shot eight people, killing four. "Griscom fought so many gun duels he cannot remember them all." One of the victims who survived is unnamed, but, per Buchanan, doesn't hold a grudge. In the light of 2020, this sounds like an officer who should at least be investigated and pulled from street work. But each shooting is ruled justifiable—in Buchanan's book, Griscom's not a threat, just a real-life Dirty Harry.

She goes into detail about two of Griscom's shootings: a 20-year-old in a stolen car from Michigan who flashed a gun at Griscom when he put his hands up, and another in a failed traffic stop that turned into a high-speed chase that evolved into an ambush with a gunman trying to shoot him. The man from Michigan died; he was wanted by police in Michigan but Buchanan never says for what or why. In the second case, the wounded driver was "an escapee from prison where he was serving life for murder and bank robbery." The two people aren't named, their families aren't quoted; according to Buchanan, they're just bad guys who did crimes. Buchanan doesn't attribute any of what she wrote here, or supply evidence to explain how she knows these narratives are true. It's implied that the main source is Griscom, for she's glowing about him as a cop, and the supplemental details from investigating agencies one would expect are either absent from her account, or woven in without attribution.

This to me is professional misconduct that goes far beyond beat capture, the necessary if ugly chumminess of making the journalism sausage. Maybe her account is true, and maybe it isn't. The dead can't speak. Buchanan knows that.

—

THE BOOK MARCHES ON TOWARD ITS MOST IMPORTANT CHAPTER, "MCDUFFIE," where Buchanan's doggedness and inherent limitations both reach their natural conclusions.

Buchanan gets a tip that police have beaten a motorcyclist to near death. Works the tip hard. She talks to sources in the medical examiner's office, who confirm he has died, and to McDuffie's grieving mother, who begs for answers. She checks out the motorcycle and can tell, immediately, that the damage wasn't from a crash. She checks the names of the officers involved: Several of them leap out, because she'd seen them a year earlier in a *Herald* series about police brutality. The night before, she remembers, she'd gotten a call from a cop. "The cop said a supervisor, aware that [Michael] Watts had difficulty in dealing with blacks, had deliberately transferred him to the predominantly black Central District." The phrase turned itself over and over in my mind after I read it—"Difficulty in dealing with blacks"—about as cruel a euphemism for racism as I could imagine.

Buchanan does what all great police reporters are expected to do. She gets the story of McDuffie's death at the hands of police onto the front page, although she concedes it is not a strong story, which she blames on an "inexperienced editor" stripping key details from it. But she got it out there, and it was enough. A day later, some of the officers involved were relieved with pay. A few others flipped, agreeing to testify. Within days, four officers would be charged with manslaughter, and later a fifth was charged with being an accessory after the fact. The case was moved to Tampa, where the judge dismissed charges against one of the accused. The other four were acquitted by an all-white jury.

The McDuffie uprising followed. Eighteen people, both Black and white, died in the ensuing days. Buildings went up in flames. The National Guard was called in.

For pages, Buchanan breathlessly details the violence, the destruction, the fear the uprising instilled in people in Dade County—especially white people. But once again, the why remains elusive. Why are so many people angry? What could be done to eliminate the cause of so much pain? Instead, Buchanan concludes the chapter by exonerating herself.

"I am not to blame," she writes. "It wasn't me who got caught up in an adrenaline-crazed chase. I didn't kill Arthur McDuffie or lie to cover it up. I was only the bad-news messenger, the reporter who found out and wrote the story. I still think about it. If it happened again, what would I do differently? I still don't know."

Earlier in the chapter, Buchanan reports that a colleague told her that during the rioting they heard an emergency room nurse shout, "You can thank Edna Buchanan for this!" Perhaps that is true. But writing is making choices. Buchanan chose to include this slight that, if true, was said by someone in an extremely stressful environment, and use it as a pivot to herself, to discuss her own role in the story. She never bothers to consider, let alone state plainly, that the likeliest answer is the most obvious one: Arthur McDuffie, a father, former Marine, and beloved local insurance salesman, should still be alive.

The account ends with Buchanan checking in on almost everyone—the cops who were acquitted, including several who landed back in law enforcement, the widow of a person killed in the uprising, even the lawyer who had represented the McDuffie family. But she never checks back in with the McDuffie family to ask how they are; there is a passing mention of a civil settlement. Their usefulness to her is done. Their loop is never closed. Years later, in an interview, Buchanan said, "Blacks will tell you that race was a factor, but I don't think so."

—

THE STORY OF A SUCCESSFUL WHITE LADY WHO JUST CAN'T STOP WRITING about murder doesn't seem shocking or revolutionary now. It feels cliché. But in the 1980s, none of the clichés existed yet. What existed was Edna Buchanan.

There's an easy allure to her story. A woman in the still male-dominated world of the 1970s and '80s, covering a male-dominated beat. She didn't have an Ivy League degree, a private university degree, or any college degree. She hailed from Paterson, N.J., where she'd worked wiring switchboards, moved to Miami Beach on a whim with her mother, applied for a journalism job only after a fellow student in a creative writing class suggested it, and spent five years working at a long-gone community newspaper before finagling her way onto the *Herald* staff—famously, by offering to write the obits.

Buchanan's fame came late, which added to the appeal. When Trillin profiled her, she was in her 40s, twice divorced, fond of dressing in slacks and silk skirts, and doing the most uncool thing a woman of any era can do—living alone with a lot of cats.

Buchanan took a leave of absence from the *Herald* in 1988 to write crime novels and never went back, although she still wrote for the paper from time to time. That same year, she appeared on *Late Night with David Letterman.* It's a strange watch. Letterman has no idea what to make of her, and Buchanan is perfectly herself. At this point, she has her talking points down pat. She recites one of her most famous sentences—"Gary Robinson died hungry"—from memory.

I spotted my own copy of *The Corpse Had a Familiar Face* in a used book bin at the Miami Book Fair, at a time when it felt like I had no choice but to buy it. From 2005 to 2013, I worked at the *Herald,* mostly the night shift, aka the night cops shift. My job entailed a mix of reporting any news that broke at night, mostly crime, as well as doing whatever the morning shift people had decided wasn't worth their time, which was mostly crime. Saying I was doing the same job as Buchanan would be a stretch. She got to work during the day, I mostly didn't. Because I was the last reporter left, I also had to cover random city meetings, sudden court verdicts, and short, quick features to go with the day's photos. In newspaper lingo, I was never "lead cops." It was more that I lurked in her shadow. I bought her book because I felt obliged to, as a fellow *Herald* reporter.

A classmate in journalism school once told me that she aspired to be a police reporter, just like Buchanan; Buchanan's work was included in our textbook collection of great works of literary journalism. When I interned at the *Herald* in the summer of 2003, Buchanan wrote a 1A story for the paper about a former central Florida officer convicted of raping and murdering an 11-year-old girl. The editors at our weekly intern meeting downtown mentioned this to the group in tones conveying that this was important, and we should all be very impressed. Her name would come up every now and then in the newsroom and at journalism conferences, always as an exemplar of what a crime reporter should aspire to be.

Buchanan's description of the grind of being a newspaper reporter contains a wealth of accurate detail that describes the job when I worked at it perfectly, and probably still does. She works weird hours. She has very little social life. She spends much of her days dealing with people who do not want to see her: cranky cops who don't want to talk, witnesses who don't want to talk, grieving families who might talk, if they can handle it. Nobody likes a police reporter, she wrote, and that's true, although it's nothing personal. Because seeing a police reporter means this is very likely the worst day of your life. She keeps her purse clean, even if everything else is a mess, so that if something bad happens to her she won't be dead with a dirty purse. I did that, too.

"I had no clue that a newspaper will swallow up your life until little is left for a novel, great or otherwise," she wrote, true then and now.

But whenever I revisit what I wrote from that time in my own life I'm struck, not by my writing or my reporting, but by how much sadness I shoveled into the paper day after day after day. People generally have two reactions to hearing you were a crime reporter: *Oooohhhh, that's so fascinating tell me everything!* Or *ooooohhhhh that's so sad tell me nothing.* The older I get, the more I wonder why nobody ever asked why my job was *necessary,* why everyone believes we live in a world filled with crime (we do not) and that it must include crime reporters.

—

A DOCUMENTARY ABOUT THE MCDUFFIE UPRISING, *WHEN LIBERTY BURNS*, came out earlier this year. Produced by Femi Folami-Browne and Dudley Alexis, who also directed, the film recounts McDuffie's life, death, and what followed within the context of Miami's decades of oppressing the Black community. Historians weigh in on the many ways Miami branded itself as a tropical, cosmopolitan paradise while also upholding segregation, forcing Black police officers to sue for representation from their own union, and paving over the thriving Black community in Overtown to expand an interstate. With the story of the Black community in Miami in the foreground, the true tragedy of McDuffie—that his death and its terrible aftermath were needless and avoidable, but for Miami's racism—is finally made clear.

Edna Buchanan's Miami is a paradise, cosmopolitan and international, with endless miles of sandy beaches, palm trees, and too many sunny days to count. Here is Miami, with its corruption, police brutality, and oppression—and don't they make for such wild stories. Making a world, or even a Miami, where these cruelties might fade away is never mentioned; there is no discussion of why the state has no income tax, and one of the stingiest unemployment programs in the nation, or why or how Miami housing remains deeply segregated. It's crime writing, sure. I'd argue it's also the ballad of white Miami, of white America, where stories of the people at the bottom are spun into tales of heroism and tragedy, while they are kept, by force, on the bottom.

But what still defines success in many a newsroom is getting something sexy. Editors love sexy. Sometimes they'll ask you for it: *Gimme something sexy, something I can sell in the meeting.* And you know who has lots of sexy stories? Cops. Newsrooms, themselves institutions with reputations to protect, can't help rewarding those in good with local police departments.

So I worry journalists will still be rewarded for writing, "Gary Robinson died hungry." I don't know they will be rewarded for asking why Robinson was hungry, or whether he should have died; he was shot dead by a security guard. Buchanan left that part out of her legendary lede.

As journalists grapple with the question of how to write about the police—to help create, perhaps, a better future—I checked to see when Buchanan, who's now 81, last wrote for the *Herald*. It was an op-ed in 2016, scolding the media for how it had covered the presidential election. Buchanan wrote that she had intended to vote for Donald Trump, although she never said outright if she did or not. For once the woman known for her crackling details chose to be elusive.

ORIGINALLY PUBLISHED IN *POPULA*, JULY 2020

THE TRUE CRIME JUNKIES AND THE CURIOUS CASE OF A MISSING HUSBAND

BY RF JURJEVICS

"My husband (since last Sat) . . . is now a missing person," read the Facebook post. "I can't believe the love of my life, my soulmate isn't here holding me."

It was July 17, 2019, and Tatiana Badra was frantic. In a series of posts, she recounted that 30-year-old Ethan Rendlen, her partner of four years, had been driving them back after a few errands to their home in The Colony, a suburb in Dallas, Texas. Rendlen, she said, had pulled over, jumped out of the car, and abruptly vanished three days ago.

Badra canvassed the businesses at the intersection. At each, patrons and staff told her that they had seen Rendlen, and that he was searching for her. After waiting by the car for hours, Badra eventually drove home alone. She had been pleading for help on Facebook ever since, posting on the pages of local Texas news affiliates and national missing persons groups.

"Police think he ate the Adderall bottle," Badra's Facebook post continued. "But why?" Badra, who claimed to be four months pregnant, said that Rendlen had just been offered a lucrative job, and that the pair were closing on an 87-acre property with a house and lake. "I just wish people would help me find the . . . father of our child. He deserves his life back!!!"

In the accompanying photos, Rendlen is tall and trim, with sandy hair and a gentle smile. In his eyes is a look of calm, of peace. Badra, a petite strawberry blonde, either cuddles lovingly beside him or mugs for the camera, working her angles and making liberal use of duck face.

Melania Boninsegna, a co-moderator of a true crime Facebook group, found Badra's post shortly after it went up. She had been searching the phrase "missing person" on the social-networking site. A 28-year-old stay-at-home mom, Boninsegna had started the True Crime Junkies in 2018, along with a co-administrator (who wished to remain anonymous in this story). It served as a private discussion group requiring permission to join, and now has 12,000 members. As new cases emerged, Boninsegna and her co-administrator created private subgroups, each linking back to the True Crime Junkies hub.

After reading Badra's post, Boninsegna sent Badra a message to see if she could help. Badra responded, both to thank Boninsegna and to share her fears about Rendlen's safety. "Girl, I won't lie," she wrote. "I'm about to lose it. I just can't stop imagining awful shit and crying." On July 23, 2019, Boninsegna started a Facebook subgroup dedicated to the case: True Crime Junkies-Ethan Rendlen-Case discussion.

Back in the days of Court TV and the O. J. Simpson trial, this kind of civilian involvement in a potential criminal case—particularly the general public communicating with a victim's family—would have required much more effort. But social media has turned viewers into users whose attention and help is often welcomed by friends and family of victims (especially when cases are solved via social media, as with the 2004 murder of Deborah Deans). The True Crime Junkies Facebook group is one of many places in a vast digital landscape—including the Websleuths site, which launched in 1999, and Reddit's r/TrueCrime—where thousands of like-minded crime enthusiasts can gather to dissect the finer points of, for example, blood evidence. The "ripped from the headlines" style of the *Law & Order* and *CSI* franchises has brought forensic crime scene analysis into our living rooms. Discussion groups picked up where the shows left off.

It isn't uncommon for civilians to do legwork for lower-profile cases in advance of law enforcement, according to Boninsegna. Texas locals, some of them members of missing-persons Facebook groups, covered a wide swath of the Dallas suburbs with missing-person flyers. While "going real life"—pestering victims' family members for updates or visiting their homes to gather "evidence"—was forbidden by the True Crime Junkies, posting flyers was a noninvasive way for members to get involved. "Every case that we have followed that had an adult male, they're kind of just put on the back burner. Women are different, and if they're a mom, then they get a lot of media attention," Boninsegna said. "I believe that most police officers don't take missing men particularly seriously."

This certainly seemed to be the case with Rendlen, at least

according to Badra. The Colony Police Department wasn't "doing shit," she wrote in a July 18, 2019, post on Dallas's NBC affiliate Facebook page; "they say bcs [*sic*] he has his wallet on his pants they'd call them [*sic*] if they found him."

But in the days following Badra's first post, things took a dark and strange turn. Her claims began to morph. She first characterized her last interaction with Rendlen in the car as a normal conversation, and later wrote that Rendlen had a "small psych fit" of nonsensical ramblings. Badra couldn't remember where they had stopped just before Rendlen ran off, and the prior destinations she mentioned kept changing: Whole Foods, a Mexican restaurant, and a nature preserve more than 250 miles from The Colony. Drugs weren't involved, then Badra claimed Rendlen had been on a "bender."

The shifting stories made many True Crime Junkies suspicious. They began to scrutinize every detail: Badra's alleged accidental melatonin overdose (impossible, many said) after Rendlen disappeared; the pregnancy claim; her recent marriage to Rendlen, which Rendlen's mother, Laura, told the True Crime Junkies had not happened. "Something isn't right with this lady," one member commented. "Too many things don't add up."

As activity in the Rendlen True Crime subgroup—which today has 5.3k members—gained steam, locals who had encountered the couple before Rendlen's disappearance began to join, too. Xaviera Crockett, then a clothing-store manager in Plano, posted a photo her employees had taken of Rendlen and Badra just hours before his disappearance on July 14, 2019. Badra's strange behavior had employees on alert, said Crockett. She'd entered the store barefoot, with "her nipple hanging out of her wedding dress," and wandered in and out of the changing rooms, which she left a mess, in just her bra. One of the employees took the photo after Rendlen threw away a bottle of clonidine, a blood pressure medication, declaring, "I won't need this anymore."

Members of the subgroup started speculating about what had happened to Rendlen. Some thought he had fled and was hiding out, others blamed Badra. "She killed him" was a frequent comment. People flocked to the threads; choruses of "any update?" followed. Moderators cautioned the group to respect the Rendlens' privacy and not to go "real life." But word had spread. Badra's Instagram posts were flooded with comments. "WHERE IS ETHAN?" "What did you do to him??"

—

PHIL LAFAYETTE, RENDLEN'S BEST FRIEND, WAS ALSO WONDERING WHAT HAD transpired between Rendlen and Badra before his disappearance. For years, LaFayette had watched as his smart and savvy friend fell further under Badra's influence. Rendlen and LaFayette, both science-minded kids, grew up across the street from one another in Glen Ellyn, Illinois, and Rendlen went on to earn a degree in chemical engineering from the University of Illinois in 2014. "He always wanted to be a chemist," LaFayette said. "To him, it was the closest thing to doing magic, to being a wizard."

Rendlen's relationship with Badra had worried LaFayette from the start. At first, Rendlen was unusually cagey about his new girlfriend, and then what he did say about her was concerning. Badra was a successful molecular biologist from a wealthy Brazilian family, Rendlen first told LaFayette, but he later revealed that she sought drugs by moving from ER to ER to avoid detection, and bought research chemicals off of the dark web. LaFayette would later find out that the two had been introduced by a former friend who allegedly met Badra during a stay at an inpatient psychiatric facility.

LaFayette was confounded by Badra's ability to manipulate his intelligent best friend. While she professed to have a large inheritance, and would treat Rendlen to fancy meals and new electronics, Badra was often in financial crisis and relayed fantastical stories about familial strife involving political corruption that she said prevented her from accessing her funds. Rendlen's father, Jeffrey, recalls that Badra said she would gain control of her funds when she turned 31, which was in September 2017. "And then the line was, 'oh, I'm not really 31, I'm really two years younger.'" Jeffrey Rendlen said he asked his son, "You're buying this?"

The couple was evicted from an apartment in 2018, and eviction notices had been filed for their Texas apartment at the time of Rendlen's disappearance in 2019. Badra had also, according to LaFayette, allegedly borrowed $10,000 from Rendlen, who had filed for Chapter 7 bankruptcy shortly before his disappearance.

Rendlen's family found that Badra's stories ranged from implausible to impossible. "Determining fact from fiction was legitimately difficult," said Chelsea Rendlen, Ethan's sister. "There was a kernel of truth in everything. It was so hard to make heads or tails of any of it." But Rendlen seemed to believe everything. During a pause in their relationship, Badra alleged that she had been kidnapped by masked men who mutilated her and left her for dead. The story grew more outlandish from there.

Rendlen told LaFayette that the kidnappers had forcibly cut the unborn child from Badra's body in a deserted cornfield. Chelsea and Laura Rendlen got a slightly different version of events from Badra. "She said she had lost the baby, but that it had cured her cancer," Chelsea recalled. Laura added, "I was told that she was pregnant with Ethan's child, and got taken out—slave-traded—to a farmhouse, that they stabbed her and so she lost the baby. But the stem cells cured her." None of them believed these stories. "It just gets to the point where it's not about reason or facts anymore," said Chelsea of her brother's relationship. Rendlen told LaFayette that he loved Badra and thought he could "save" her.

It was a maddening situation for the family. Rendlen was unable to see what those who loved him found obvious: he was being conned. But, like any good scam, Badra's had begun with developing a powerful psychological hold over Rendlen. Those mechanics of manipulation don't "happen overnight," said Alexandra Stein, a visiting research fellow at London South Bank University who specializes in the study of cults and dangerous social relationships. "This is a process. You get the initial come-on, which is very nice and flattering, and creates rapport and starts building trust."

Badra's seeming generosity with her inherited money, coupled with constant tales of distress, made for a persuasive lure: a "love-bomb," in which the victim is showered with attention, affection, and sometimes gifts. Rendlen's appointment as a knight in shining armor to her constant distress was the clincher. Badra also kept Rendlen isolated from his family, managing to convince him that he had been molested as a child by a relative. "Scammers work with fear," Stein explained. "A corollary relatedness is urgency: 'if you don't help me now, I'm going to lose my child, my house, my life.' And also the threat that you might lose a relationship that purports to be beneficial to you, but is actually causing you chronic stress. That creates a trauma bond. All of these things work to prevent you [from] using your systematic thinking."

Family and friends were hopeful when Rendlen—who had mostly held short-term positions as a quality technician and geotechnical engineer—landed his dream job as a chemical engineer after he and Badra moved to Texas in 2018. "He was like, 'I finally made it, bro,'" LaFayette said. "He had his own cubicle, they gave him his own company credit card. His bosses were coming to him with projects to work on. They wanted his direct input and he was so excited about that."

Then Rendlen called LaFayette with the news that he and Badra were moving to Florida. The details of the plan didn't seem to track. Badra, with no prior experience, was planning to set up a real estate business, backed by a mysterious uncle who suddenly wired her inheritance payments. Rendlen's burgeoning career would be left behind. "This is the part that really got me because it was so unlike Ethan," LaFayette remembered. "He told me he was going to drop everything and he was going to try and be a crab fisherman. I said, 'Ethan, that's crazy . . .' The fact that he would be willing to drop his childhood dream to be a crab fisherman? It was insane. This girl is telling him crazy things, and he's just eating it up. Something's gonna happen; something bad is gonna happen."

That phone conversation was less than a week before Rendlen disappeared—one of several strange calls that had his family and LaFayette concerned and confused. A few days before he had gone missing, Rendlen called his father and told him that he had been robbed at knifepoint in his apartment and needed emergency money. After Rendlen had disappeared, Badra called Laura Rendlen in tears. Rendlen had wanted to call off the wedding, she said. This was news to the family, who had not heard about any plans to marry. (Though, according to Rendlen's past relationship status updates on Facebook, the pair had already married in 2018.)

Now, 900 miles away in Illinois, LaFayette's mind was spinning. Something bad had happened. His best friend was missing, and nothing about the circumstances made any sense. The primary source of information was Badra—until Rendlen's family and LaFayette joined the True Crime Junkies group. Rendlen, LaFayette would learn, had not been Badra's first mark.

—

IN JUST 24 HOURS AFTER ITS CREATION, THE RENDLEN TRUE CRIME JUNKIES group was buzzing with information. Most of it was about Badra.

Members and admins posted their finds in rapid-fire succession. One was an archived GoFundMe from 2013 that Badra had allegedly launched to pay for various medical bills. She listed seizures, bone marrow issues, cancer, and a blood-clotting disorder as her diagnoses, and raised $1,100 of her $7,000 goal, according to the archived page. Users surfaced multiple social media profiles with photos of Badra under a variety of names (many of them derivatives of her legal name), and several hints at pregnancy Badra had made on Instagram in 2017 and 2018, not to be mentioned again. "In all of the posts, not one, have I seen her actually pregnant," one True Crime Junkies member commented. "What is happening to these babies (if she is actually pregnant)?" Several members also discovered that photos Badra had included in a Facebook post of the 87-acre house she and Rendlen were allegedly purchasing were of a $55 million dollar home in Florida's affluent Gables Estates.

Alice, who did not want their real name used, was watching as this unfolded. Alice had known Badra not as Tatiana, but as "Anya," a creator of online support groups for Brazilian survivors of sexual assault and eating disorders. Badra was also, Alice contends, a ruthless cyberbully. From 2007 through 2012, Badra formed secret groups in which she would routinely leak nude photos of friends and acquaintances, and instigate online fights. In all of these groups, Badra would solicit money for medical treatments that she seemed not to have undergone, according to Alice and multiple sources who knew her during this period.

On July 24, 2019, Alice made a post in the True Crime Junkies and other Facebook groups alleging that Badra had trashed the Chicago apartment they shared briefly in 2012, did not seem to work or attend school, and told elaborate lies. "I have received, over the years, messages of her boyfriends . . . and others that came after . . . telling me how they felt victimized bye [*sic*] her, how she scammed them of THOUSANDS of dollars, got them hooked on drugs, was hooked on drugs, faked pregnancies, faked suicide attempts, etc etc," Alice posted to the Facebook group. "She's not a cancer survivor . . . she's currently an American resident because of her fake story of being a survive [*sic*] of abuse, she's dangerous, abusive and manipulative."

After Alice spoke out, the dam broke. Screenshots from other social media platforms surfaced, posted by the True Crime Junkies members, and more of Badra's former acquaintances began commenting in post threads about their own alleged experiences. Many comments were from Badra's ex-boyfriends, and followed a similar pattern of dubious pregnancies and medical conditions, chronic drug use, threats, manipulation, and stolen money.

"I went from thinking, 'my friend Anya has big boyfriend problems' to 'all of Anya's boyfriends have big Anya problems,'" Edward Grabowski, a former friend of Badra's, wrote in a comment thread. Grabowski had known Badra for about a year between 2012 and 2013, during which time he gave her money, bought her a phone, and helped her after chemotherapy treatments that he said turned out to be fabrications. Grabowski alleged that Badra had been heavily abusing Norco (a pain reliever) and Klonopin (a benzodiazepine), and routinely used hospitals as her sources for these drugs.

While most of her ex-associates had known Badra as Anya, she'd also gone by Tatiana Nikolaevna, Pippa Althofen, Tatiana Lyubov, Tatiana Lebedeva, Aniia Lilya, Lilja L., Stazia K., and Lavinia Badra.

In ten years, 34-year-old Badra may have been at least ten different people. As Anya, she was a chemotherapy patient. As Aniia, she engaged in discussions on DNA and racial "purity" on the white supremacist website Stormfront. Tatiana Lebedeva was a radical feminist; Pippa Althofen a sugar baby.

Badra also appears to have been involved with identity fraud. Documents obtained by *Vice* indicate that she has been associated with at least six different Social Security numbers. Some belong to other people entirely, according to database records. "In my 30+ years of conducting tens of thousands of background investigations, I have rarely seen this many SSNs linked with a single subject," wrote the private investigator who reviewed the documents for *Vice,* via email. "I can't definitively say that she is committing fraud, but the fact that she is associated with so many . . . unexplained SSNs seems to lend credence to the fact that it could be for fraudulent purposes."

Most of these identities shared a history of dramatic claims: tragic beginnings as an orphan in radiation-riddled Chernobyl, then her adoption by a wealthy Brazilian family into an elite, pampered childhood marred by various forms of abuse. Badra claimed to have genius abilities in science and music, which brought her to study first in London and then the United States. Then there was the inheritance, with a spiteful uncle presiding over her funds. Badra also said she had been a model, a cancer survivor, and the mother of a young daughter back in Brazil. At various points, she claimed she was working at high-paying jobs as a financial analyst, a senior sales engineer, and multilingual translator. She was about to become a doctor.

Very little of this would turn out to be true. Chernobyl, cancer, and her work history were all allegedly fictions, as several of her former friends had confirmed years ago after speaking with Badra's adoptive mother. None of the schools she claimed to have degrees from—Northwestern University, University of São Paulo, and the University of Texas, Dallas—have records of Badra having attended. While Badra was adopted at birth, she, according to several sources who allegedly confirmed this with Badra's mother years ago, was not born in Ukraine. Former close friends say Badra had no signs of the significant scarring and other physical trauma that would have resulted from the forced removal of a fetus via amateur C-section (just an appendectomy-like scar above her right hip of about two inches, according to one ex-boyfriend). And as the True Crime Junkies had suspected, Badra and Ethan Rendlen were never legally married.

When con artists start out online, it begins "a grooming process to actually desensitize you to some of the things that come after that," said Martina Dove, author of *The Psychology of Fraud, Persuasion and Scam Techniques*. "By the time you are asked for money, or asked to believe something that's ludicrous, you're invested. You know something's wrong, but you just can't pull back."

The internet was Badra's home base for a reason: in an online relationship of any kind, intimacy is built quickly. She was able to be anyone on social media with little effort, forging connections online before transitioning them to in-person meetups. As soon as those around her became more than casually suspicious, Badra would be on to the next identity, and her next set of marks, leaving the shells of her former selves behind in abandoned accounts, purged blogs, and a handful of avatars.

—

WHILE THE INTRICACIES AND INTIMACIES OF A SCAM ARE WHAT HOOKS A VIC-tim, these are also the elements that simultaneously hook us: the readers, the viewers, the writers. Scams have all the markers of a good drama—mystery, suspense, plot twists, and bad guys.

Con artists give us a complex sort of villain, an antihero: even if a con artist is a wholly unsympathetic character, there's titillation to be found in their gumption (and, in some cases, outright genius). Scammers, according to Alaleh Kamran, a Los Angeles–based criminal defense attorney with 30 years of experience, "are smart enough to have succeeded in any area," but "one-upping the system—there's a thrill to that." Even more compelling is the razor-thin line between brazen and foolish, which works out better for some than others, particularly those preying on less-sympathetic victims.

Anna Sorokin, better known as Anna Delvey, rose to notoriety, if not outright fame, after bilking socialites, celebrities, the Beekman hotel, and other bastions of luxury out of $275,000 by pretending to be the heiress to a $70 million trust fund. Her story fascinated the public and news outlets dubbed 2018—the year multiple stories about Sorokin broke—"The Summer of Scam." This bizarre twist on underdog popularity led to an Anna Delvey episode of HBO's *Generation Hustle,* followed by a life rights deal with Shonda Rhimes for her upcoming series *Inventing Anna,* starring Julia Garner. Even her victims did well; Rachel DeLoache Williams, a former photo editor whom Sorokin stuck with a $62,000 hotel bill, wrote a tell-all that made *Time*'s 100 best books list for 2019. Sorokin, who served just under four years in prison, doesn't seem at all derailed by her life of crime. She was paid $320,000 for her story ($45,000 more than she stole) and seems to be enjoying life on the outside—at least it appears so on her Instagram.*

But as much as we love to watch someone like Sorokin buck the system and—all things considered—win, focusing on the moment when a scheme absolutely fails can be even more of a thrill. This setup is at the heart of shows like ABC's *The Con,* which premiered in October 2020 to 2.6 million viewers and is narrated by Whoopi Goldberg. The more spectacular the scam, the harder the fall; in Episode 5, we learn how 50-year-old Anthony Gignac, raised in Michigan, managed to convince the richest in Miami that he was a Saudi prince by affecting an accent and a flashy style. Gleefully, Goldberg details the undoing of this ruse: a replica diplomatic license plate Gignac affixed to his car, which he bought online for $79.

There might be a hint of gleeful schadenfreude when scammers like Sorokin or Gignac scam the ultra-rich, using outrageous tactics to do so. But there are, of course, much more sobering cases of deception.

—

ON JULY 23, 2019, PHIL LAFAYETTE AND LAURA RENDLEN ARRIVED IN TEXAS, determined to find something that would lead them to Ethan Rendlen. Nearly two weeks had gone by since his disappearance, and there had still been no communication from him. They spent most of their time driving from location to location, trying to find a

* Anna Sorokin was jailed again for seventeen months for overstaying her visa. On October 5, 2022, Sorokin was released on bond, and is on house arrest pending deportation.

match based on Badra's versions of events. Five days later, the two left Texas to return to their jobs, without a conclusive answer.

It took four Dallas locals—all civilians, who had connected via a now-defunct Facebook group on Rendlen's case—to put it together. One of them was Amber (who did not want her last name used), then a substitute teacher and off for the summer. Amber was able to spend several days in her car, driving to any location on her GPS that matched Badra's limited descriptions. On July 26, she was able to locate the car wash, the gas station, and the bar where Rendlen had last been seen and verify, with others, that Badra had been there. Another civilian helping with the search—who did not wish to be identified—spoke to several people who had seen the couple. "The car wash guy said, 'yeah, there was this couple here, they were fighting, [and] he ran off that way towards the woods,'" Amber said. "'They were both high as a kite. She runs around looking for him, trying to find him, basically all night. And they can't find each other.'"

The group located the embankment, behind a gas station, that Badra had mentioned, and notified the Dallas police. Responding officers did not see any direct evidence linking Rendlen to the area, but passed the information on to the Colony Police Department. The officers urged the search team not to return to the area, which they said was an extremely dangerous hotspot for drug activity.

On July 29, 12 days after Rendlen had disappeared, the Colony police located his body in the embankment. The case was then turned over to the Dallas Police Department. "The girlfriend stated that her [*sic*] and the comp [Rendlen] were in the area of Rosemeade Pwky [*sic*] and Marsh Lane using drugs on 7/14/19," reads page 10 of the Dallas Police incident data sheet report for Rendlen's case, obtained by *Vice*. "The girlfriend states that her [*sic*] and the comp had gotten into a verbal dispute and he got out of the vehicle and left walking in an unk [*sic*] direction." In the report, Badra seemed to have a better recollection of their last known location than she had previously admitted.

The last Chelsea and Laura Rendlen saw of Badra was on August 3, 2019, when they went to gather Rendlen's personal items from his and Badra's apartment. That day, Badra, wearing a red wig, biked up to the apartment complex, and the building manager, who denied Badra access to the apartment due to the eviction notice, notified the Rendlens. "The manager said she'd been going around burying Ethan's stuff all weekend," Laura recalled. They called the police, and watched from the street as Badra was arrested on four open warrants for traffic violations (no other charges seem to have been filed against her).

This was Badra's only arrest during this time period, and, according to police records, she spent less than an hour in police custody. "We have heard from one of the detectives we are working with that she was released into the custody of a police officer," said Chelsea. "And that's kind of where that part of the information ended."

Sergeant Jay Goodson of the Colony Police Department and Detective Guy Curtis of the Dallas Police Department, both of whom were assigned to the case in their respective departments, did not respond to repeated interview requests. There was no indication from the police report provided to *Vice* whether or not Badra was ever under investigation by law enforcement.

Rendlen's death was ultimately ruled as accidental/unknown by the medical examiner, whose final sign-off on the autopsy was dated October 30, 2019. "Part of the problem is because of decomposition, and the length of time, even the medical examiner said that drowning can't be ruled out," said Anita Zannin, a forensic scientist, who reviewed documents pertaining to Rendlen's case for *Vice*. "Once the organs start autolysing—turning to mush, essentially—it's harder to make those determinations. Their hands are kind of tied when the medical examiner comes up with 'accident' as manner of death."

After Rendlen's body was found, Boninsegna and her True Crime Junkies admin team received private messages from more victims, many of whom wished to remain anonymous. Some alleged they had been coerced into providing Badra with money, others that they had been blackmailed into purchasing items for Badra, who had threatened to make false allegations against them to police.

Francis Silva, Badra's ex-husband, thought she was likely connected with Rendlen's death. His relationship with Badra, which began online in 2006, resulted in disaster. Silva alleges he discovered she was lying about a cancer diagnosis and threatened to divorce her, to which Badra responded with a false domestic violence claim against him in order to obtain a green card via asylum. On July 10, 2012, Badra sent him an email in which she confessed to having fabricated the abuse. "So be it, I lied about what happened," she wrote. "I perjured myself. I was angry, I was scared . . . I LIED AND I TAKE FULL RESPONSIBILITY." Silva did not respond. In August 2012, Badra rescinded the charges, for which Silva had been indicted by the Texas district attorney at a grand jury trial in 2011.

While there have been a battery of accusations leveled against Badra by her former friends and romantic partners—lies, theft, coercion, physical abuse—she has never been formally charged with any of these crimes in either Illinois or Texas. None of the ex-associates who spoke with *Vice* have brought charges. "Yes, I did rebuild my life away from her," Silva said of his experience with Badra. "I try my very best to forget that ever happened to me."

Badra declined an interview request for this article. "I apologize, but no," she wrote via email, calling the allegations against her "Absolut [*sic*] insane, all of it." She did not deny involvement in Rendlen's death, which she described as a "tragic accident" and "the worst trauma I carry with me, for many reasons." Badra was adamant that the online discourse surrounding the case had been particularly hurtful and damaging after such a loss. "I could spend forever talking about him and how much I miss him and love him, and what type of person he was," Badra wrote. "These people, these 'sleuths,' have caused me enough grief for enough lifetimes already."

When *Vice* reached out to Badra again for comment on this article, she denied using multiple Social Security numbers, stating: "I've obviously never used anyone else's SSN other than my own." In a follow-up email, she declined to comment any further. "I have retained legal counsel and have been advised by my attorney to not make any statements to you. You and your editor should be hearing from them soon," she wrote. We never heard from legal counsel on Badra's behalf.

—

IN THE MONTHS AFTER RENDLEN'S DEATH, BITS AND PIECES OF INFORMATION about Badra surfaced sporadically in posts in the True Crime Junkies group—an arrest for D.U.I.; photos of an Amazon package addressed to her old Texas apartment; a stay at an ayahuasca retreat. Behind the scenes, Boninsegna and her moderation team received more private messages from those who had encountered Badra.

In August 2020, longtime friends Davis Trent, then 26 years old, and Tiffany Harris, who was 25 years old, came forward with harrowing accounts of having met Badra, known to them as 28-year-old Anya Audi (Badra was 34 at the time). They had learned Badra's true identity from a misplaced medical form, and a Google search led them to the True Crime Junkies group. When Trent and Harris called the Colony Police Department with this information, they said they were told that Badra was "dangerous" and to change the locks to their apartment.

Trent claims that, while he was under the influence of ketamine, Badra convinced him that he had been molested by a family member (as she had with Rendlen), and that she played him interviews with serial killers like Ed Kemper and 911 calls of rapes in progress. "She comes up with these outlandish, horrifying lies about people, then plants them in your brain while you're tripping," Trent said. "And you've got to understand that she didn't just say things. She has done her research. She knew terminology that she could use to make you think it was real." Harris also alleged that Badra dosed her with methamphetamine, and then psychologically manipulated her.

But even more peculiar is that both Trent and Harris relay Badra's recounting of Rendlen's death. On several occasions, they said she broke down completely, bursting into uncontrollable bouts of tears to confess that she had witnessed his last moments. "I watched him die," they said she would say, over and over. "I watched Ethan die."

As of this writing, Badra uses the name Tatiana on Facebook Dating, where she claims to be 29 years old, and Tanya (a diminutive for Tatiana) on Instagram. On both platforms, she has claimed Jewish familial lineage, despite her past activity on Stormfront and her Catholic upbringing (an event program obtained by *Vice* lists Badra as completing her first Holy Communion in 1997). Occasionally, Badra will post about Rendlen. Davis Trent found her Reddit account still logged in on his computer (the account was also sent to *Vice* by another independent source and deleted after *Vice* reached out for comment). "I was present when my fiancé had a psychotic break and made a run for it," Badra wrote in r/eyeblech, a subreddit dedicated to gore and postmortem photographs. "Some absolute psychopaths on Facebook gave me the gift of spamming my email with his autopsy photos." (The Dallas Police Department's Open Records Unit was able to confirm the release of autopsy records, but not the requestor's identity.)

Even recently, Badra's life, as she recounts it, is filled with stories of high drama and suffering. Members of the True Crime Junkies group posted screenshots of Badra celebrating the sixth month of a pregnancy on her Instagram account, a claim that most members suspected was untrue. But on March 3, Badra posted a photo of her newborn daughter, born 19 months after Rendlen went missing, to her Instagram.

The True Crime Junkies started buzzing again. "I believe it is her baby and hope that being a mother at last for real will make her change her ways," one member commented. Others were less optimistic. "Oh snap!" another member wrote. "I was team 'she's faking this pregnancy!'"

A few weeks later, Badra wrote a sobering post about her baby's hospitalization for seizures. "Still no answers to the 'why' of the epilepsy," it read. "We have an appointment with genetics on Monday to go over the epilepsy gene panel, then neuropeds on Wednesday. Send good vibes her way!"

ORIGINALLY PUBLISHED IN *VICE*, AUGUST 2021

HAS REALITY CAUGHT UP TO THE "MURDER POLICE"?

BY LARA BAZELON

At 7:45 p.m. on December 27, 1986, Faheem Ali was shot dead in the streets of Baltimore. No physical evidence tied anyone to the killing, and no eyewitnesses immediately came forward. But Baltimore homicide detectives Thomas Pellegrini, Richard Fahlteich, and Oscar "The Bunk" Requer were not going to give up easily.

Requer was later immortalized as a central character in David Simon's iconic HBO series *The Wire*. As Simon wrote in the afterword for his acclaimed 1991 nonfiction book *Homicide: A Year on the Killing Streets,* Requer "lives on in Wendell Pierce's portrayal of the legendary Bunk Moreland on *The Wire,* right down to the ubiquitous cigar." Pellegrini, meanwhile, was the jumping-off point for Detective Tim Bayliss, a character in the long-running television show *Homicide: Life on the Street,* which was inspired by Simon's book. Requer and Pellegrini are among a constellation of Baltimore Police Department officers who have, through Simon's work, defined what it means to be a homicide detective in the popular imagination—and whose biggest cases are starting to fall apart or have been overturned.

Determined to find out who killed Faheem Ali, Pellegrini, Fahlteich, and Requer honed in on 12-year-old Otis Robinson, who was outside when the shooting happened. They allegedly brought Robinson and his mother to the police station and separated them, questioning the seventh-grader alone. Robinson told the detectives that when he left his house to go to the corner store, he saw a few men across the street in conversation, though he didn't notice much in the dark. As he continued walking toward the store, he heard a gunshot and fled.

Even though Robinson insisted he could not identify a shooter, the detectives showed him an array of photos, including one of Gary Washington, a 25-year-old Black man, according to a lawsuit Washington filed against the city and the detectives in 2019. Robinson knew Washington, but he made clear that he did not see who shot Ali. The detectives wrote down this statement.

Then, according to the lawsuit, the questioning took a turn. "Cooperate," the detectives allegedly told the 12-year-old, "or you'll never see your mother again." Unless Robinson identified the shooter, the officers allegedly continued, he could be charged with homicide.

Robinson "crumbled under the pressure" of threats from the detectives, according to the lawsuit, and signed a second statement falsely identifying Washington as the shooter. His first statement was never turned over to prosecutors or defense attorneys for Washington. (Attorneys for the defendants have denied liability in court pleadings but declined to comment, stating that they were "constrained to speak only through the judicial process.")

Five months later, Pellegrini testified at a pretrial hearing. The lawsuit says he "committed perjury" by telling the court that Robinson was not threatened or coerced when he implicated Washington. The next day, Washington was tried on first-degree murder and weapons charges. On the witness stand, Robinson testified that Washington was the shooter. Washington's attorneys called multiple witnesses who testified that the killer was a man named Lawrence Thomas, but the jury believed Robinson. As a judge later wrote, "For all intents and purposes, Otis Robinson was the state's entire case." Washington was convicted of Ali's murder and sentenced to life in prison.

Robinson recanted his testimony in 1996 to an investigator for Washington. He did the same in court in 1999 and again in 2017, explaining he had been strong-armed by detectives. In 2018, a judge overturned Washington's conviction. In 2019, the Baltimore City State's Attorney's Office dismissed the charges against him. Lauren Lipscomb, the deputy state's attorney who oversees both the Conviction Integrity Unit and Police Integrity Unit, stated, "We respect the finding of the judge who found Robinson's recantation credible. Evidence insufficiency is not the same as factual innocence and evidence insufficiency is the reason we dismissed."

Washington, now 57, walked free. He spent more than three decades in prison. Whether the detectives who put him there will face any repercussions remains to be seen.

—

SINCE 1989, 25 MEN CONVICTED OF MURDER IN BALTIMORE HAVE BEEN EXONERated, according to the National Registry of Exonerations. Official misconduct was present in 22 of the cases. "The history of BPD officers and detectives withholding exculpatory evidence from the accused, coercing and threatening witnesses, fabricating evidence, and intentionally failing to conduct meaningful investigations is decades long," wrote the attorneys for Clarence Shipley, a Baltimore man who spent 27 years in prison for a murder he did not commit before he was exonerated in 2018. "BPD's misconduct in [Shipley's] case," they said, is "yet another chapter in the long story of BPD's pattern and practice of wrongdoing during homicide investigations."

Baltimore homicide detectives have coerced witnesses (including children), fabricated evidence, ignored alternative suspects, and buried all of that information deep in their files, attorneys for Washington and other exonerees say. "So much of this is a war mentality that is infused with a strong racist edge," said Michele Nethercott, who retired in July 2021 as the director of the University of Baltimore Innocence Project Clinic. "It is a war out here and we just do whatever we have to do and if that means threatening kids and threatening witnesses, we will do it. They use the same tactics on the witnesses as they do on the suspects."

More than a dozen such cases can be traced directly to misconduct by the Baltimore Police Department in the 1980s and 1990s. Many of the detectives accused of being bad actors—Pellegrini, Requer, Fahlteich, Donald Kincaid, Gary Dunnigan, Terrence

McLarney, Jay Landsman, and several others—were profiled in Simon's book *Homicide*. Some of them, like Pellegrini, Landsman, and Requer, inspired beloved television characters on *Homicide: Life on the Street* or, later, *The Wire*.

On *The Wire,* Bunk's supervisor was "Jay Landsman," just as in real life Requer's boss was Detective Sergeant Jay Landsman. Simon says the character John Munch in *Homicide: Life on the Street* was "a combination" of Dave Brown and Terry McLarney (though a caption in the 2006 edition of *Homicide* identifies Munch as being based on Landsman, too). McLarney has also been accused of misconduct. Brown, who died in 2013, is implicated in the suppression of evidence in Shipley's lawsuit, though he is not named as a defendant. The lawsuit says that Brown and others "acted with impunity in an environment in which they were not adequately supervised" by McLarney and Landsman.

These men became bold-faced names in *Homicide,* Simon's chronicle of the year he spent with their elite unit in 1988. The suspects say little other than to issue blanket denials or outright lies. And the dead of course cannot speak. So it is the "murder police"—an expletive-spewing, gallows-humored brotherhood—who take center stage. In Simon's telling, they are stubborn, hard-drinking, and at times highly unpleasant. But the reader also comes to believe that they are dogged in their pursuit of justice.

"[Y]ou are one of thirty-six investigators entrusted with the pursuit of the most extraordinary crimes: the theft of a human life," Simon writes. "You speak for the dead. You avenge those lost to the world. Your paycheck may come from fiscal services, but goddammit, after six beers you can pretty much convince yourself that you work for the Lord himself."

Read through the lens of what we now know about the criminal legal system, however, the book reveals a dark side to this God complex: an arrogance, overreach, and ends-justify-the-means mentality that resulted in wrongful convictions and ruined lives. The excruciating pressure Otis Robinson said that detectives used on him is on florid display.

In one scene, Simon describes how Pellegrini and Landsman attempt to solve the killing of a man named Roy Johnson. Potential witnesses are brought in, including a girl wearing a yellow miniskirt. Looking her over, Pellegrini thinks, *Helluva body, too.* When she refuses to cooperate, Landsman screams at her, "YOU'RE A LYING BITCH."

Berating her fails to produce results, but Landsman continues, "You just got a charge, you lying piece of shit." Then, as he leaves her alone in a cramped interrogation room, he calls out to Pellegrini, "NEUTRON THIS BITCH." This is merely a request for a swab of her hands, but Simon writes that "Landsman wants to leave her stewing on it, hoping she's in that box imagining that someone's about to irradiate her until she glows." It's just one example of "the Landsman blitzkrieg," which Simon tells the reader "often succeeded simply because of its speed."

Detectives from *Homicide* worked on the cases of at least six of the 25 men exonerated for murder who are identified in the National Registry. One man, James Owens, was convicted in 1988 for the murder of a young woman and served 20 years in prison before he was exonerated by DNA evidence. James Thompson Jr., the state's star witness against Owens, was interrogated multiple times by Pellegrini, his supervisor Landsman, and Detective Dunnigan. Each time, Thompson told a different story. The final version came after hours of coercion by detectives "to force him to fabricate a story," according to a lawsuit filed by Owens. In 2018, Baltimore officials settled with Owens for $9 million, the largest settlement in the city's history.

To date, legal settlements related to the Baltimore homicide unit have cost Maryland taxpayers at least $45 million. Eight exonerees have pending federal civil rights lawsuits. The detectives deny all wrongdoing, and their lawyers declined to comment for this story.

The list of wrongful convictions is growing: On December 21, Paul Madison's conviction was vacated by a Baltimore City Circuit Court judge after he spent 30 years in prison for a murder he did not commit. Early this month, Baltimore announced an $8 million settlement to the family of Malcolm Bryant, who served 17 years for a murder he did not commit and died in 2017 at the age of 42.

Other cases that are not counted as exonerations by the National Registry raise similar concerns about Baltimore's homicide detectives. Among them is Wendell Griffin, convicted of the shooting death of James William Wise in 1982. In a federal civil rights lawsuit filed in 2013, Griffin accused homicide detectives Kincaid and Brown, along with Landsman's brother, Lieutenant Jerry Landsman Sr., of suppressing witness statements "excluding Mr. Griffin as the shooter." In 2012, when public-records requests revealed that exculpatory evidence had never been shared with the prosecution or defense, Griffin accepted a plea to time served rather than face a retrial—and it was on that basis that his federal lawsuit was dismissed. While the state's attorney's office has previously cited that the known evidence supports Griffin's guilty plea, Griffin is still fighting to clear his name. Lipscomb stated, "Our position is that it is a resolved case and I have no further comment."

The Baltimore Police Department is not an outlier. In recent years, similarly ingrained misconduct has been revealed at police departments in New York, Boston, Philadelphia, Kansas City, and Chicago. What is different is the veneer of gritty integrity that has long burnished the images of Baltimore's homicide detectives.

"Overall, following those detectives on other cases from January 1988 to December 1988, I did not see police work in which evidence was purposely mishandled or in which exculpatory evidence was purposely ignored or obscured," Simon wrote in an email to *New York*. "That may be because as a civilian, I didn't recognize such moments, or because my presence during casework made such behavior prohibitive. I can't say."

—

***HOMICIDE* RECOUNTS INTERROGATION METHODS BY DETECTIVES THAT FEW** police departments would endorse today. These brutal, dehumanizing tactics come across as an ugly but necessary strategy to secure convictions. In *Homicide,* Simon writes that, in 1988, the Baltimore police's murder clearance rate was 74 percent—in 2020, by contrast, the department's clearance rate was 40.3 percent—and the reader is given no reason to believe that the numbers represent anything other than the guilty getting caught. If anything, Simon wrote at the time, Baltimore juries convicted too seldomly. "In truth," he writes, "juries want to doubt." As a result, "the chances of putting the wrong man in prison become minimal." (In an email, Simon wrote that he would now reconsider his skepticism about the likelihood of wrongful convictions: "I minimized the chance of an investigative or prosecutorial error—never mind purposed misconduct—making it all the way to a jury and conviction; that chance is more substantial than I once believed.")

Three decades later, the portrait of policing in *Homicide* lands differently. Maryland has just over 6 million residents, but in 2019 it ranked sixth in exonerations, tying with Florida, which has a population of nearly 22 million. Only Illinois, Pennsylvania, Texas, New York, Michigan, and California had higher totals. We are regularly informed of high-profile exonerations in Baltimore and elsewhere, including the men wrongly convicted of killing Malcolm X; Anthony Broadwater, cleared in the 1981 rape of best-selling author Alice Sebold; and Kevin Strickland, freed in November after spending 43 years in prison for a triple murder in Kansas City, Missouri, he did not commit.

But when *Homicide* was published in 1991, DNA testing was in its infancy, the Innocence Project did not exist, and wrongful convictions were viewed by many as a fever dream. In 1993, U.S. Supreme Court justice Antonin Scalia, roundly rejecting the argument that a prisoner could bring an appeal based solely on his innocence, wrote, "With any luck, we shall avoid ever having to face this embarrassing question again."

In the criminal legal system, treating child suspects and witnesses like hardened adults was also a common practice, even though children are easily intimidated and vulnerable to coercion, and therefore likely to say whatever police want. The 1990s was the era of the "super-predator," a term coined by the Princeton political scientist John J. Dilulio. Super-predators were conscienceless child criminals who roamed the streets committing rape, murder, and mayhem. To protect society—and especially their own communities—they had to be locked away.

Consider the case of Ransom Watkins, Alfred Chestnut, and Andrew Stewart Jr., all 16 when they were convicted in Baltimore in 1984 for the shooting death of 14-year-old DeWitt Duckett in his junior high school so they could steal his jacket. Watkins, Chestnut, and Stewart were sentenced to serve life terms in an adult prison.

Midway through *Homicide,* a dramatic confrontation takes place between Watkins and Donald Kincaid, the Baltimore homicide detective who brought him down. It is the summer of 1988, and the Baltimore police have been enlisted to investigate a riot at the Maryland Penitentiary, where Watkins is serving his life sentence. Landsman and Kincaid set up shop in the deputy warden's office, questioning a steady stream of shackled suspects. Most of the prisoners decline to speak with them, some less politely than others.

Watkins—described by Simon then as "a thick-framed nineteen-year-old monster"—has something to say, and it is not about the riot. Staring down Kincaid, the teenager asks, "How the hell do you sleep at night?" To which Kincaid replies, "I sleep pretty good. How do you sleep?"

Watkins retorts, "How do you think I sleep? How do I sleep when you put me here for something I didn't do?" Angry and on the verge of tears, Watkins continues, "You lied then and you lyin' now." Kincaid responds that Watkins is guilty. The teenager tries to argue, but Landsman cuts him off, calling for the guards to take him away. "We're done with this asshole," he says.

In November 2019, a Baltimore judge found Watkins, Chestnut, and Stewart factually innocent after the state's attorney's office conceded they were wrongfully convicted.

The case against Watkins, Chestnut, and Stewart—known as "the Harlem Park Three"—turned on the purported eyewitness testimony of four middle schoolers. All have since recanted, claiming that they testified falsely under relentless threats and pressure from Kincaid and his partners John Barrick and Bryn Joyce. One of the witnesses, Ron Bishop, recently told *The New Yorker* that "if

I didn't tell them who did it, I could be charged with accessory to murder." Bishop, 14 at the time, grew desperate: "I was thinking, *Should I get a gun and blow my brains out?* I was torn between committing suicide or, you know, go into court and tell these bunch of lies."

Lipscomb called the detectives' conduct in the Harlem Park Three case "appalling," and said that "what was even more troubling was that they were putting these juvenile witnesses in a patrol car and taking them to [the police station] without their parents after they had given interviews to detectives with their parents in their homes. So there is one story when the parents are present and it appears that the detectives were not happy with that story and so they took the teenagers down to homicide." According to Lipscomb, one witness, now an adult, heard his mother screaming outside the interview room demanding to know why police had taken her son without her knowledge.

Collectively, the Harlem Park Three served 108 years in prison. In March 2020, the State of Maryland awarded the three men nearly $9 million. In August 2020, they filed a federal civil rights lawsuit seeking unspecified damages for the violation of their civil rights. Attorneys for the three alleged that Kincaid's investigation was sloppy and biased—Kincaid himself testified that he took no notes during the interviews. And according to the state's attorney's office, Kincaid told Watkins, "You have two things against you—you're Black and I have a badge." (Kincaid has denied all wrongdoing.) "This triple exoneration," their lawyers wrote, "is the largest wrongful conviction case in American history."

While the Harlem Park Three grew into middle-aged men behind bars, Michael Willis, the alleged murderer, went free. This, too, is attributed to the detectives' misconduct. As Kincaid and his team pursued Watkins, Chestnut, and Stewart, they had evidence implicating Willis in Duckett's murder, including witness statements that Willis threw away a gun and wore the victim's jacket, all on the day of the murder. (Willis was murdered in 2002.) None of that information was turned over to the defense, as required by the 1963 U.S. Supreme Court ruling *Brady v. Maryland.*

Simon says that he accurately reported the encounter between Watkins and Kincaid. He also wrote, "I didn't cover any exonerated cases in 1988, the year I was permitted in the homicide unit. So my first-hand knowledge of the casework in question is nil."

Watkins remembers the encounter differently, telling *New York* that Kincaid wanted him to provide information on the prison riot and that he didn't even recognize Kincaid at first. He also disputed Simon's physical description of him, stating that he was not particularly big at the time. "I think the word 'monster,' frankly of all the people we've seen come in and out, I would not call him a monster," said Lipscomb. "This is a soft-spoken, good-natured person."

—

JUST LAST MONTH, LANDSMAN, MADE FAMOUS IN *HOMICIDE* FOR HIS ALL-CAPS battering-ram interrogation method, became a defendant in yet another wrongful conviction lawsuit. In a complaint filed in federal court on December 14, Shipley, the Baltimore man who served 27 years for a murder he did not commit, alleges that Landsman, McLarney, Robert Bowman, and Richard James—all depicted in *Homicide*—violated his rights and caused his wrongful conviction in the 1991 murder of Kevin Smith.

The complaint alleges that Landsman and McLarney, another

squad leader in *Homicide,* failed to supervise when at least one detective hit 18-year-old Allan Scott over the head, chained him to a chair in the interrogation room, and refused him medical treatment for hours. The lawsuit alleges Scott gave false testimony in exchange for leniency related to pending theft charges. Shipley's lawsuit also alleges that the detectives buried evidence implicating the real killer, Larry Davis, who died in 2005.

The lawsuit includes a photograph of a handwritten note from a Baltimore police employee to homicide detective David John Brown memorializing an October 26, 1991, phone call with the victim's brother, Edward Smith. The note reads, "shooter is Larry Davis." According to the lawsuit, "by the time trial began, the Officer Defendants had changed Edward Smith's story from implicating Larry Davis to implicating Clarence Shipley. To secure the conviction of Mr. Shipley, the defendants made sure that key evidence, such as the note above, was not provided to Mr. Shipley and his lawyer. As a consequence, when Edward Smith took the stand and pointed the blame at Mr. Shipley, Mr. Shipley's lawyer had no meaningful way to show that the morning after the murder, Mr. Smith had implicated Larry Davis—not Clarence Shipley."

Homicide, now three decades old, is very much a product of a particular time and place in the annals of American criminal law. Nethercott, the former head of the University of Baltimore Innocence Project Clinic, says it is also "a cautionary tale for embedded journalism." In the foreword to a 2006 edition of *Homicide,* writer and longtime Simon collaborator Richard Price addressed this critique: "Are writers like us, writers who are obsessed with chronicling in fact and fiction the minutiae of life in the urban trenches of America, writers who are in fact dependent in large part on the noblesse of the cops to see what we have to see, are we (oh shit . . .) police buffs?"

Price determines they aren't, and Simon points out that his next book, *The Corner,* takes the point of view of those "being policed and hunted" during the height of the war on drugs. And *The Wire* provides a kaleidoscope of perspectives from beautifully drawn characters, including cops who are blatantly violent and racist, which is central to why the show was groundbreaking and beloved by so many. "I believe in writing from the point of view of characters as a function of embedded narrative," Simon said. "This doesn't mean you don't include the bad with the good, or change outcomes, but it does demand that you do your job and deliver the worldview of your protagonists for all to see." Simon said that in both *Homicide* and *The Corner,* "the same process of empathetic embedding was employed regardless of where I stood."

Still, Price's question is worth considering. As Price noted, the kind of intimacy necessarily created by this kind of prolonged and up-close exposure gives rise to "an unavoidable empathy" for the writer's subjects. It can also lead to a story that allows some characters the full complexity of three dimensions while flattening others, depriving them of their humanity and readers of the full story.

The detectives of *Homicide,* for their part, seem to have long been comfortable with Simon's reporting. In an afterword to the 2006 edition, Simon wrote that they "requested remarkably few changes" when he showed them the manuscript. McLarney, who was later promoted to lieutenant, wrote in an addendum that he and his colleagues were "gratified" by Simon's portrayal.

Nor do the detectives appear to have significant regrets about their careers. Jay Landsman retired in 1994 to work as a law enforcement officer for the county, where he was promoted to lieutenant in 2015. Reflecting back on his time in the Baltimore Police Department's homicide unit, he told the *Baltimore Sun,* "I never had a bad day down there, I loved it."

ORIGINALLY PUBLISHED ON THE CUT, IN PARTNERSHIP WITH THE GARRISON PROJECT, JANUARY 2022, AS "DAVID SIMON MADE BALTIMORE DETECTIVES FAMOUS. NOW THEIR CASES ARE FALLING APART."

PART III

SHARDS OF JUSTICE

WILL YOU EVER CHANGE?

BY AMELIA SCHONBEK

Late on a fall afternoon ten years ago, Cheryl and Troy walked into a room and shook hands. It was a small space at the Justice Center in Portland, Oregon, almost entirely taken up by a conference table and chairs. Beads of rain covered the room's one long window. Cheryl sat next to it so she could look out, which helped remind her to breathe. She had barely eaten that day, just enough so she wouldn't be sick to her stomach.

Cheryl and Troy were strangers, though, in one sense, they knew each other well. For years, Cheryl had been in a string of violent relationships, and Troy had a long history of getting drunk and abusing his partners. In 2005, he went to prison for 22 months for choking his girlfriend. Cheryl and Troy met that afternoon because both of them wanted desperately to change, yet nothing had freed them from the destructive patterns they were in. By this time, Cheryl, who was then in her 60s, had tried therapy and found it isolating to sit opposite someone who hadn't lived through violence. And Troy, then in his 40s, attended Alcoholics Anonymous, though he sometimes struggled to accept the pain he had caused others without making excuses. After years of trying to move on from their experiences, they both discovered restorative justice, a form of conflict resolution that brings together survivors and offenders with a focus on repairing the damage done, rather than punishing the person responsible. They each agreed to participate in a practice called a surrogate dialogue.

The question of how to respond to incidents of domestic and sexual violence has never had a particularly good answer. The criminal justice system has long been the only option, and the tiny number of people whose cases even made it to court had to choose between reliving their trauma and not seeking justice at all. In 2017, when Me Too broke open the collective rage and grief that had been building inside survivors after decades of being dismissed and disparaged, the question became impossible to ignore. The more these stories—which fell on a whole spectrum of abuse, from workplace creepiness to rape—were told, the clearer it became that the existing options for resolving the instances were seriously limited. The accused were called out, a few were convicted of crimes, and some were fired or divorced. Then what? "If we want the #MeToo movement to be about more than just which celebrity will be the next to fall, or whose comeback must be stopped—if we want it to lead to real, lasting, and widespread cultural change—we need to talk," wrote the journalist Katie J.M. Baker, "about what we do with the bad men."

A lot of people weren't ready to have that complicated conversation right away. It was thrilling to be able to speak out about the experience of harm and feel heard. It was possible, finally, to see men like Larry Nassar and Bill Cosby, who'd assaulted dozens of women and girls over years and years, convicted of their crimes. Sending them to prison looked like an acknowledgment of all the pain they had caused and a warning to other men that they couldn't get away with abuse. It felt like something to celebrate. But that warning hasn't stopped more stories from surfacing. And after being sent to prison, Cosby's conviction was overturned on a technicality—even the catharsis of seeing the most egregious cases closed didn't last. So now, nearly four years on from Me Too, we're left looking forward, trying to untangle the intricate issue of what the consequences should be for people who have caused harm, and to figure out the harder thing: how to welcome them back into society while also caring for the people they have hurt. Practitioners of restorative justice think their approach is one way to navigate it all. Their work focuses on what survivors need to recover, and the process is designed to benefit the larger community as well: If you help people understand the impact of their choices, the thinking goes, they may change how they act in the future. Though restorative justice is often used to resolve cases involving young people or low-level crimes, women's advocates are divided over whether to apply it to domestic- and sexual-violence cases. And only a handful of programs have ever done it.

When she first began to consider a surrogate dialogue, Cheryl was apprehensive. Long ago, she had learned you should never give an abuser ammunition, because he would use it against you. What if she met with a guy and then afterward he tried to track her down? Still, she had some questions she badly wanted to ask all the men who'd hurt her: *Didn't you care about me? Have you learned anything from this? What are you doing to keep from doing it again?* Cheryl didn't know whether the dialogue would be meaningful or whether she would be able to get through it. But she couldn't stop thinking about the question that had stuck with her the longest: *What did I do to make this abuse happen in my life?*

An organization called Domestic Violence Safe Dialogue (DVSD) had paired Cheryl with a mentor named Marci who would help her prepare for the exchange and be present as her advocate. Cheryl learned that the framework for the dialogue gave her complete control over the situation, the length of the meeting, how deep the questions went, and what the goals of the conversation were. Later, they planned how Cheryl could respond if her dialogue partner said something inappropriate or frightening. They set a code word that Cheryl could use if she needed Marci to speak up in her defense or stop the conversation entirely.

In the room that day, Marci sat down beside Cheryl, with Troy and his advocate across from them, and a facilitator took a seat at the head of the table. Marci conferred quietly with the facilitator while everyone else sat still and silent. Inside, Cheryl panicked. When Troy sat down, she wondered what would happen if he got mad, if he reached across the table and hit her. She remembered how she had learned to keep herself calm during performance reviews at work: She placed her hands on the table, very quietly, and looked Troy straight in the eye. Troy had walked into the room feeling calm, sure that he would be able to handle whatever came up. *If I want to stay sober,* he told himself, *I have to do this.*

—

CHERYL: I WAS WILLING TO DO THIS BECAUSE I DIDN'T WANT TO CARRY THIS FEAR, GUILT, SHAME, RESPONSIBILITY ANYMORE. I HAD DONE MY BEST TO GET RID OF IT BY MY OWN MEANS, BUT I STILL HAD IT.

—

CHERYL GREW UP IN A MIDDLE-CLASS NEIGHBORHOOD OUTSIDE PORTLAND during the 1950s and '60s in a house that overlooked fields and orchards. In the summer, she would often sleep in the backyard with friends. She knew not to invite them in because her dad could lose it at any moment. When Cheryl was a baby, her mother took her to the basement when she cried to keep her father from hitting her. Cheryl doesn't remember much about growing up, but she remembers the night her father said to her mother, "I'm going to take this gun, and I'm going to kill those kids and then kill myself." Her three brothers would run out of the house when their father was enraged, but she often stayed to try to negotiate between her parents. When Cheryl was 19, her father died. *I'm so glad,* she thought. Soon after, the whole family got together to celebrate. (To protect her identity, Cheryl asked that only her first name be used.)

After her father was gone, Cheryl's life didn't change in the ways she had hoped it would. She had a boyfriend who drank too much and hit her. When they split up, she began seeing a man she met through a coworker. Soon he began to ask her to list everyone she had talked to while they'd been apart. Cheryl was taking karate lessons, and her boyfriend didn't like all the time she spent away from him. "What's more important?" he asked. *Crap,* Cheryl thought. She felt herself being pulled back toward a familiar dark place, where the need to make a man happy blotted out everything else. She called her boyfriend and told him it wasn't going to work out. After they hung up, he came over and beat and choked her until her eardrums ruptured.

The next day, Cheryl called her mother, who called the police. "You tell him everything," Cheryl's mother instructed her when the detective arrived, and she did. The detective listened, then Cheryl remembers him saying, "We can arrest him, but the laws on this are pretty weak. You'll have to come and tell your story, and he'll say you're a liar. It might just make him more angry, and he might kill you the next time."

From the outside, it was usually hard to tell that anything was wrong in Cheryl's life. She worked at a big corporation, and people around her would say, "You're so good at your job, you're so confident." Yet, again and again, Cheryl's boyfriends turned out to be controlling or abusive, just like her father had been. *Maybe I am broken,* she thought. *Maybe all this is my fault.* Some days were so hard that she would think, *I just can't do one more day like this.* She thought about ending her life.

When she was 37, Cheryl began to see someone new. The man didn't hit her, but there were signs. "Are you going to wear that?" he'd ask. "You don't know how to make good decisions." Cheryl saw what was happening and said to herself, *You cannot do this again.* It took her a year to end the relationship. Six weeks after the breakup, the man drove up to her house, stepped from his car, and shot himself in the head in front of her. As her neighbors rushed him to the hospital, Cheryl stood outside her house and waited for the police to arrive. She had reached bottom. "I am either going to die," she realized, "or I am going to stick around and do things differently." This is how Cheryl wound up in a room with Troy, a man she didn't know who had been violent with his girlfriend, to try to understand why he had done what he did.

—

TROY: I FLY BY THE SEAT OF MY PANTS. I'LL JUMP INTO ANYTHING FEET FIRST. WHEN THEY ASKED ME TO DO THE DIALOGUE, I DIDN'T GO INTO IT LIKE, "OKAY, LET'S GET THIS OVER WITH." IT WAS, "OKAY, I BELONG HERE, THIS IS WHAT I NEED TO DO." IT WAS A RELIEF, I GUESS: I'M DOING THE RIGHT THING.

—

TROY GREW UP THE ONLY CHILD OF A SINGLE MOTHER. HE WAS POLITE, THOUGH he always liked to be the center of attention, and he hated when he made mistakes. The first time Troy tried alcohol, as a teenager, he drank so much he passed out on a neighbor's couch. The next morning, he woke up confused about how he'd gotten there. He got wasted a couple more times in high school, on each occasion drinking until he was ill. Although he hated getting sick, he loved feeling cool and making people laugh, and he wanted to drink more and more.

Troy graduated from high school and got a job in construction. It was the late '80s, and it seemed that everyone he worked with had a drug or alcohol problem. He started making a lot of money, more than he needed. It didn't really matter what substance someone put in front of him—if it was there, he was going to do it.

Troy always thought that growing up with a single mother had given him a solid respect for women. When he was drunk, though, everything went out the window. In 1992, he got married and adopted his wife's daughter. Soon they had another daughter together. He never hit his wife, but they argued all the time. Troy would pick fights to justify storming out of the house and going to drink; he wasn't around for his kids, and he withheld money from his wife.

His marriage broke up in 2001, and afterward he lived on the street. A few years later, he began dating another woman, and he fought with her, too. She hated Troy's drinking, and when she confronted him about it, he lied. During one fight, Troy began to choke her. *What am I doing?* he realized, and he ran out of the house. He made it only three or four blocks before the police caught him.

After his arrest, Troy felt that he no longer knew who he was. He felt sorry for himself, and angry that the public defender assigned to his case pushed him toward a plea deal.

Troy took it and went to prison for 22 months. His 9-year-old daughter wrote him a letter saying she must be the reason he couldn't quit drinking. It broke his heart. He joined AA and kept it up when he got out: Drug and alcohol treatment was a condition of his early release. He found 12-step work to be moving, and when he reached the fifth step and sat across from his sponsor, Troy admitted to him "the exact nature of my wrongs." Troy was used to talking about the things he had done—he'd had to recite them in court and to his probation officer—although in those cases he was thinking, *What do I want to hide so that I don't look bad?* With his sponsor, he felt for the first time that he could be truly honest. "Alcoholism does a good job of isolating, making you feel like nobody knows your problems," Troy said. When he read a list of his wrongs to his sponsor, a man who had hurt people and had made amends, Troy realized, *Oh, I'm not all alone. We're all dysfunctional.* "It took the shame away," he later said.

AA emphasizes completing actions on the path to recovery,

and Troy took the approach to heart. *Okay,* he thought, *I have to continually do something. That's my medicine for my alcoholism.* He asked the facilitator of his treatment program if she knew of any ways he could give back to the community, and she gave him a phone number for DVSD. Troy wondered if this thing called restorative justice was in some way self-serving. But what he wanted most was to stay sober, and he was willing to give anything a try.

—

CHERYL: I LIVED AFRAID FOR 38 YEARS OF MY LIFE. I DIDN'T KNOW THAT PEOPLE DIDN'T FEEL THAT WAY ALL THE TIME. IN THE DIALOGUE, THAT WOUNDED PART OF ME WAS FIGHTING ABOUT BEING IN A ROOM WITH AN ABUSER AND OPENING UP, BUT I WAS READY TO GO. IT FELT LIKE COMING TO A CRESCENDO.

TROY: HONESTLY, WHEN I HEARD ABOUT THE DIALOGUE, I THOUGHT, *IF THIS IS GOING TO HELP KEEP ME SOBER, I'LL DO WHATEVER IT TAKES.* TO SIT ACROSS FROM ANOTHER INDIVIDUAL AND DO THIS—IF IT HELPS THEM, AND IN TURN HELPS ME, THEN I'M OKAY WITH THAT.

—

RESTORATIVE JUSTICE IS A MODERN INTERPRETATION OF A VERY OLD IDEA: ITS central principles resemble the conflict-resolution methods of Navajo and Maori people, among others. Its contemporary Western form can be traced back to one spring night in 1974, when two teenage boys slashed tires and broke windows all through a Mennonite community in Ontario. A court officer named Mark Yantzi was handed the case. After the boys pleaded guilty, Yantzi noticed that the legal system offered no mechanism for them to literally repay their debt to the people whose property they'd damaged.

Yantzi proposed to the judge that they all try something new: The boys would knock on the door of each victim's house and ask what the damage had cost them. A few weeks later, the boys would return, this time with money to repair the houses or cover insurance costs. One of the homes they visited was owned by a woman who had lived in it nearly all her life. Before, she had never locked her doors or felt unsafe. The house had a picture window, and the boys had thrown a brick through it. She told them that after the vandalism, she became so fearful that she couldn't be at home alone. "But now I sit across from you," she said to the boys, "and I see that I really had no reason to be afraid." "You got a sense that they were making connections . . . that helped to kind of bind them back into the community," Yantzi later said.

Yantzi's early experiments unfolded alongside a growing awareness among progressive-minded activists that sending people to prison is a dehumanizing practice that answers violence with more violence, rather than with rehabilitation. As they searched for more holistic, less punitive ways to resolve criminal cases, community groups and allies in the legal system picked up Yantzi's work and added to it.

Nearly five decades later, restorative justice encompasses many interconnected ways of resolving conflict, both within the traditional justice system and outside it. The idea, for example, of atoning for the destructive violence of slavery through reparations is rooted in restorative principles. Forty-five states allow some form of restorative justice into their criminal proceedings; a judge can divert a case into a restorative process—so that a young or first-time offender might avoid a conviction, for instance—or a victim's family may ask for their case to be resolved this way. In other instances, those involved in a crime might turn to nonprofit restorative-justice programs long afterward to bring closure beyond what the courts could offer. And in cases of domestic and sexual violence, such processes could be a potential way of seeking justice in the many situations where prosecutors decline to bring charges.

The fact that restorative justice was an option available to Cheryl as a survivor of domestic violence was deeply controversial even ten years ago. In the '70s, while Yantzi was looking for ways to resolve cases outside the courtroom, mainstream feminist activists were trying to get the legal system to take domestic violence and sexual assault seriously. The system had long told women who were abused that it was a private matter to be worked out with their boyfriends or husbands. Many activists thought that enacting laws against domestic and sexual violence would help survivors find justice and dissuade men from harming women. So they fought to pass legislation requiring arrests in domestic-violence incidents, criminalizing marital rape, and making it easier for women to take out restraining orders against men who harmed them.

The problem was this strategy mostly didn't work. In 1995, the year after the Violence Against Women Act was signed into law, the National Institute of Justice noted that there was "little conclusive evidence" that criminalizing domestic violence actually discouraged offenders or protected victims. Even now, as more and more people admit that legal solutions alone aren't enough, the question of whether restorative justice is appropriate in cases of rape and domestic abuse has not gone away.

"More police and more jails are never going to solve the problem," says Beth E. Richie, who leads the Department of Criminology, Law, and Justice at the University of Illinois, Chicago. "There was never any illusion that we could arrest our way out of this thing." As an anti-violence activist in the 1980s, Richie saw how cops used force with Black communities: It mirrored the violence many women experienced from their partners. She and her fellow organizers instead advocated for another approach, one that acknowledged patriarchal abuses of power. They did not want to simply remove offenders from their communities. They wanted their communities to work with offenders and help them figure out how to stop being violent, and they wanted women who had been abused to be able to express what they needed. In a lot of cases, it turned out, what those women wanted was not to be questioned by detectives or cross-examined in open court but to talk about what had happened on their own terms.

—

CHERYL: THE MAJORITY OF WHAT I WANTED WAS TO BE HEARD. AND I WANTED TO HEAR THE TRUTH, AND I DIDN'T KNOW IF HE COULD TELL ME HIS REAL TRUTH OR NOT. I WANTED TO ASK, "WHAT DO YOU BELIEVE SET YOU OFF?" I COULD TELL HE DIDN'T KNOW HOW TO TALK TO ME ABOUT WHAT HAD HAPPENED. BUT I JUST WENT FOR IT. I THOUGHT, *THIS IS MY TIME TO ASK THOSE QUESTIONS THAT HAVE BEEN CONTROLLING ME.*

TROY: I AM FORTUNATE—OR UNFORTUNATE—THAT I NEVER BLACKED OUT WHEN I WAS DRINKING. I KNOW ALL THE CRAPPY THINGS I DID AND HOW I TREATED PEOPLE. WHEN IT CAME TIME, I DIDN'T LEAVE ANYTHING OUT.

—

CARRIE OUTHIER BANKS, WHO FOUNDED THE DOMESTIC VIOLENCE SAFE DIA-logue program Cheryl and Troy participated in, began working at a women's crisis shelter in the 1990s. She soon realized that all her clients blamed themselves for how their partners treated them, no matter what anybody else said. She saw how it kept many women from leaving violent relationships: "'If I had just tidied the kids' shoes,' they'd say, 'maybe he wouldn't have hit me.'"

Outhier Banks had read about a program in Canada that arranged conversations, called surrogate victim-offender dialogues, between unrelated rape survivors and perpetrators who had taken responsibility for their actions. The surrogate approach was experimental, intended to meet the needs of victims for whom a direct dialogue with those who had harmed them was impossible. "The survivors were able to ask the perpetrators the question that everybody asks," Outhier Banks told me. "'What did I do wrong?' And the perpetrators were able to say, 'Nothing. You were just in the wrong place at the wrong time.' That's exactly what the survivors I worked with need to hear: 'It wasn't your fault.'"

Outhier Banks finished a Ph.D. in conflict resolution and became one of only a few facilitators who were willing to use restorative dialogues in cases of domestic violence. At the time, most feminist advocates were highly skeptical. They felt that the particular power dynamics of gender violence made it near impossible for victims and offenders to engage in any kind of balanced conversation.

"Men can't change—that's a lot of what I heard," Outhier Banks remembered. Through the early aughts, when she told stories about her work to other domestic- and sexual-violence-survivor advocates, a lot of people got angry. "We were hated," she said. Some feminist advocates responded that supporting restorative justice meant you cared more about perpetrators than about victims. Others told her they were concerned that survivors couldn't handle sitting across from batterers. "I was like, 'Really? I think these women are pretty strong.'"

Outhier Banks was under no illusions that the approach was right for everyone. She had friends who had left violent relationships who told her they had no interest in meeting an offender for a dialogue. "I don't need some man telling me he was wrong," they said. "Of course he was wrong."

"Justice means different things to different people," Outhier Banks said. "You can never fully have justice, right? Something's been taken. For me, it's about what we can do to make a victim as whole as possible." The survivors she worked with generally had clear and unique ideas about what they wanted out of the process. One woman told Outhier Banks that all she wanted was to walk into the room, face someone who had committed violence that was similar to what her ex had done, and be strong enough to walk out. Another was ready to end her dialogue incredibly fast, as soon as the man she was paired with admitted to wrongdoing. "I wanted to hear someone take responsibility because my ex never will," she said. "I just needed to know that there are men out there who can change."

DVSD's dialogues didn't always go smoothly. Sometimes men said they would be able to take responsibility for their actions, and when the time came, they couldn't. Outhier Banks remembers a dialogue in which a man didn't own up to a lot of what he had done to his wife, and the woman on the other side of the table called him out. He tried again to admit the ways he'd manipulated and hurt his ex. Often, the man said, when his ex came into the room and started a conversation, he would turn around and leave while she was mid-sentence because he knew it got under her skin. The woman paused. "Wait, what?" she asked. "Did your ex ask you if you did that on purpose?" She did ask, the man said, and he told her she was crazy every time. That was familiar to the woman: Her own ex-husband had done the same thing. "You aren't crazy," the man assured her. "He did that on purpose, and I can tell you that because I did it, too. Because it keeps you off your footing, it keeps you in the relationship." Outhier Banks remembers that, afterward, the woman kept saying, "The mind games . . . I knew I wasn't crazy."

Though there are now hundreds of restorative-justice programs across the country, only a handful work with victims of domestic and sexual violence. Even established programs like DVSD have struggled to overcome resistance to the approach: After losing crucial funding in 2019, it stopped offering surrogate dialogues altogether. Because each restorative-justice program uses slightly different methods, it's hard to assess how well they work—and make the case for their existence—beyond participants' feedback. What research there is points to their value. A 2014 study found 49 percent fewer cases of post-traumatic-stress symptoms in crime victims who went through a dialogue process compared with those who did not. And recent studies have routinely found that offenders who participate in restorative processes are less likely to be rearrested than those who move through the traditional justice system.

Underneath all the back-and-forth about whether, and when, to use restorative justice, there are bigger, murkier questions: If a man has been violent in one relationship, or several, will he ever be capable of acting differently? Whose responsibility is it to accompany him through the process of interrogating his thoughts and actions? What do we do about the men who are unwilling or unable? And if a man is able to meaningfully change, and the woman he abused doesn't want to have anything to do with him, how should the community respond?

By the time Me Too unfolded, Sonya Shah had been thinking about these questions for a decade. Shah facilitates surrogate dialogues in the Bay Area for those who have committed and survived sexual harm. Like Outhier Banks, she has often been asked to justify her focus on the rehabilitation of bad men. "Understandably, everybody is so pissed at how little attention gender violence has gotten, at how much victim blaming and silence and shaming there has been," she told me. "That needs to happen. After hundreds of years of silence, it's okay to be angry." Shah is herself a survivor of sexual assault. She long ago realized that her healing could not be dependent on another person's suffering. "It is very difficult to hold that we can be furious, but that doesn't mean that people who have committed this harm are irredeemable," she said. In Shah's work as a facilitator, keeping both of these ideas in mind at once is critical. "For men to be able to break the silence of their most horrible secrets, they have to believe you're not going to judge them," she said.

If a person has beaten or raped someone, moving past judgment is among the hardest things to do. What Shah and other restorative-justice practitioners advocate requires a huge amount of trust—from individuals and society—that men who are given a second chance won't take advantage of it. It makes sense that survivors and their supporters would hesitate to offer a potential out to someone whose motivations may be self-serving—say, reducing a prison sentence—rather than sincere. And when punishment is framed as closure, as a way to acknowledge and value a victim's pain, it might feel impossible to turn down. "We go to the punitive because we don't know there's any alternative," Shah told me.

Revenge is also exhilarating. The desire to see a person who has inflicted violence suffer as a result is very old and very human. It's present in progressive circles—where the public shaming of those accused of wrongdoing is routine—just as it is among law-and-order conservatives. In the years following Me Too, Shah has often thought about the impulse to pathologize the perpetrators of sexual and domestic violence, to label them monsters. "Do we do that because they're everywhere? Because they're our fathers, our coworkers?" she wonders. A common refrain in conversations about gender violence is the need to reform the toxic culture that allows men to physically and sexually assault women with little consequence. If one way into that cultural shift requires moving past judgment and extending hope for change to the bad men, will we be willing?

—

CHERYL: I TALK TO PEOPLE WHO SAY, "PEOPLE WHO BREAK THE LAW JUST NEED TO BE PUNISHED UNTIL THEY FIGURE OUT THEY'RE NOT SMARTER THAN THE REST OF US." WHEN THEY SAY THAT, I HEAR THE FEAR IN THEIR VOICES THAT THE WORLD IS OUT OF THEIR CONTROL AND THEY DON'T FEEL SAFE IN IT. LORD, THAT WAS ME FOR A LONG TIME. WE DO THE SAME THING WE KNOW UNTIL WE KNOW SOMETHING DIFFERENT.

TROY: I WAS TOLD NUMEROUS TIMES IT'S VERY IMPORTANT TO FORGIVE YOURSELF. I TEND TO HAVE LOW SELF-ESTEEM. MAYBE I'M NOT WORTHY OF BEING FORGIVEN. BUT AT SOME POINT IN TIME, I ACTUALLY STARTED LOOKING AT THE WORK I'VE DONE. THAT WAS WHEN IT STARTED CLICKING. *OKAY, YOU'RE NOT OLD TROY. THIS IS THE NEW GUY.* I'M NOT THE A-HOLE THAT MY HEAD TELLS ME I AM.

—

THE ROAD TO GETTING A SURVIVOR AND AN OFFENDER TOGETHER FOR A DIAlogue is usually long and uncertain. Over the course of a decade, DVSD facilitated only around 200 of them. Though the process has a lot in common with therapy, it's more narrowly focused, and ideally the two work in tandem. Most people who haven't been through it "have a very naive view of what it takes for someone to be accountable," Shah told me. "They don't understand what it actually means to look at yourself, your trauma, privilege, and gender socialization and to unpack that and change it."

Cheryl sometimes spoke to groups of men who had harmed their partners, and she started to notice that a lot of them had stories that began similarly to hers, with fathers who abused them. She began to wonder if abusers and survivors were two sides of the same coin: people who had experienced horrible things but responded to them very differently.

Troy struggled with the complexity of recognizing the circumstances that had shaped his life while also fully owning the way his actions hurt those around him. He had learned that addiction is a disease, one he surely had. And he knew that many people were alcoholics but did not choose to come home and choke their partners. "I still try to justify my actions," Troy told me. "I have to say, *Okay, this is what I did,* and leave it at that." The lasting impact of incarceration made it harder. "It keeps getting brought up and brought up and brought up," he said. "It's like, 'Hey, I'm trying to move on and be better, yet you're reminding me about a really crappy time in my life.' Even if I make amends, I'll still have this label. When is my debt to society paid?"

Restorative justice is built on the assumption that people who have done violence don't want to keep doing it, that they want to change. Even when they do, the process can be excruciating. And, of course, many people have no interest in acknowledging bad behavior or acting differently. Restorative justice will never be a solution for men such as Cosby and Weinstein who have unapologetically wielded their power to evade consequences. There's no clear path by which to rehabilitate them as long as they deny or minimize what they've done. For restorative justice to take off—and for society to figure out how to diminish the overwhelming amount of bad behavior—it will depend on the very cultural shift it hopes to create, and it will need to unfold alongside many other interventions in order for that to happen.

—

IN A COURTROOM, MANY STORIES ARE TOLD, AND THE TASK IS TO WINNOW them to a single true story: Someone is guilty of a crime, or they aren't. In a restorative-justice dialogue, every story, or history, can be true at once, even those that seem to contradict one another. Inevitably, these dialogues mean different things to each person who takes part, and each person remembers them differently.

Cheryl had been sitting across from Troy for 15 minutes when she realized that he might be just as scared as she was. The room was small, and the table and chairs where they sat took up most of it, so that everybody felt physically close. She watched Troy's expression as he talked about his relationship with his ex. She could see that he didn't know what she wanted or how exactly to tell her what he had done. She realized that Troy was there for the same reason she was: "He couldn't move on from it alone."

Troy doesn't remember feeling afraid. Because he had done so much work in AA on accountability, his preparation was quicker than most; and only a few weeks had passed between his first meeting at DVSD and his dialogue with Cheryl. When he told people about restorative justice, most of them asked him, "Why would you do that? Why would you relive what happened?" The way Troy understood it, he wasn't reliving it if he knew that talking about it would help someone else. It was a step that took him "that much farther away from potentially doing that sort of thing again."

Sitting across from Cheryl, he told the familiar story: how he began drinking as a kid, about his marriage and the arguments and the different forms of abuse, and about the night when he choked his ex-girlfriend. As he spoke, he had a feeling he knew well from his time in AA, which was that he wasn't alone in his experiences. Feeling that gave his memories less power than they'd had before. "I guess this is making my peace with it," he said later. "A kind of agitation that doesn't exist anymore after a while."

Cheryl watched Troy as he told her what had happened, and hearing about his life made her feel less alone, too. At the end of her story, she told Troy about her ex-boyfriend who had died by suicide. *Oh my gosh,* Troy remembers thinking. *She's carrying the guilt of this guy killing himself in front of her.* He wanted very much to tell her that she wasn't responsible for what had happened that night.

After an hour, Cheryl was ready to ask Troy the question she had turned over in her head for years: "When you were fighting with your ex, when you were hurting her, what were you thinking about her?" The question threaded through each of her relationships, all the way back to her father. When she asked it, Cheryl remembers that Troy looked right at her and said, "I wasn't thinking about her at all. I was just thinking about how angry I was."

Cheryl asked to take a break, and she and Marci stepped outside. It wasn't raining anymore, and the air was cool and tasted clean. "I knew at that moment that it wasn't my fault at all, the way I had been treated by all the men in my life," Cheryl said. "I had been told that over and over, but until this abuser could look me in the eye and say it, I didn't believe it."

They went back inside, and Cheryl talked to Troy for a while longer. She wanted to make sure he knew that violence "was a broken path for him," that he was doing the things he needed to do in order to make different choices. He told her he was sorry for everything she had gone through. She believed he was sincere. Then she left, feeling, for the first time, that she wasn't carrying around the entire weight of everything that had happened to her.

The dialogue didn't fix everything Troy or Cheryl struggled with in their lives. Troy was still an alcoholic, and staying in recovery was hard. It took Cheryl many more years to really figure out how to live differently with the knowledge that the abuse she'd experienced wasn't her fault. A couple of years after the dialogue, she ran into a woman she'd been acquainted with for a long time. The woman said to Cheryl, "I know you don't know this, but the man you sat with is my son." Since then, Cheryl has occasionally seen Troy when she's over at his mother's house. They sometimes sit down and talk a bit, catching up.

ORIGINALLY PUBLISHED IN

***NEW YORK MAGAZINE*, JULY 2021**

THE PRISONER-RUN RADIO STATION THAT'S REACHING MEN ON DEATH ROW

BY KERI BLAKINGER

As soon as I drive past the East Tempe Church on the outskirts of Livingston, Texas, I can hear the laugh track on my radio. It's from *Martin,* a three-decade-old television sitcom. The fictional Detroiters' racy wisecracks seem incongruous crackling through my car speakers on a winding country road.

When the laughter dies down, the slight Southern lilt of a DJ named "Megamind" cuts in to introduce the next segment.

"Bringing it to you room service–style," he says, signing off with a catchphrase that's a little bit tongue-in-cheek: Like most of his listeners, Megamind doesn't have a room. He lives on a metal bunk in a maximum-security prison, and his real name is Ramy Hozaifeh. To the men in the Allan B. Polunsky Unit, he is best known as a regular voice on 106.5 FM The Tank, the prison's own radio station.

The Tank is so low wattage you can only hear it for a minute or two after you leave the parking lot. But the programming is as plentiful and varied as any commercial station on the outside, with shows covering everything from heavy metal to self-improvement. It's all recorded in a studio hidden deep inside the prison and stocked full of equipment, most of which was donated by churches and religious groups. It doesn't have the fame or following of San Quentin's *Ear Hustle* podcast, but The Tank allows men on one of the most restrictive death rows in the country to have a voice that reaches beyond their cells. Usually—just like in most lockups—the prisoners at Polunsky are not allowed to write letters to each other. But for the radio station, the warden carved out an exception, allowing them to pass along essays and poems for the staff chaplains to deliver to Hozaifeh and his fellow DJs, affording the most isolated men in Texas a rare chance to be part of the prison community.

Every morning, Hozaifeh plays an episode of *Martin* or *Sanford and Son*—shows that still make sense for listeners who can't see the action because they're locked in a cell with no television. "You can listen to their clowns," he said. "You don't have to see them at all."

Like most lockups, life in the roughly 3,000-man prison an hour and a half north of Houston is pretty bleak, especially for the high-security prisoners who spend most of their time in solitary. That includes a few hundred men isolated because they're considered dangerous or in danger, but it also includes nearly 200 men on Texas's death row. For years, the guys on the row have been disconnected from the prison's general population. They can't go to the mess hall or the chapel or the main yard, so most of the time they only meet their fellow prisoners in passing—like when janitors come by to mop or hand out towels. They can't go to classes or prison jobs, and they don't have tablets or televisions. But they do have radios.

The first time I heard about the station was from a man on death row named John Henry Ramirez. It was a week until he was scheduled to be executed, and I'd visited him to ask about his plea for prison officials to let his Baptist pastor lay a hand on him as he died. He answered my questions about his faith and whether he feared death, but what he really wanted to tell me about was the radio station.

"When you get out to the parking lot, you can just tune in, and you'll hear," he said. By the time I got back outside, he explained, I could catch the noon news update with the day's menu. "It's become such a huge part of Polunsky," he added. "You should hear all the people talk about it."

—

THE STATION STARTED IN EARLY 2020, WHEN WARDEN DANIEL DICKERSON ARrived at Polunsky, and some prisoners approached him with a question: Would he let them start a radio station?

He'd been asked all sorts of strange questions in the 24 years he'd worked for the Texas prison system—but this one was a first. Still, he decided to hear the men out.

"When they explained it and what was going to be done—and of course everything's pre-recorded, so it can be looked at and reviewed—it didn't sound like a bad idea," he said.

In his eyes, it seemed like a radio station could help give the men something to care about and connect with—especially when the prison was too short-staffed to expand their programming any other way. And in the early days of the pandemic, Dickerson said, it also seemed like a great way to help prisoners all across the facility understand what was going on, even those who couldn't leave their cells.

"They may not all have TV, but most everybody has a radio," Dickerson told me. "And anybody who's been on a cell block knows some folks will turn the radio up loud enough where even if you didn't have one, you're probably going to hear it anyway."

The first time he sat down in his office and tuned in, he did not regret it.

"It's your own little prison city radio station," he said, flashing a cockeyed grin. "And you can walk around and see the change in people."

Even as a visitor, I can see it, too. Usually when I interview men on death row, we talk about their cases or their upcoming death dates or the conditions they live in. But now, they rattle off the programming schedule they know by heart. There's *Smooth Groove*—that's R&B—on Sundays, then rap on Mondays and Latin music on Tuesdays. There's a night for Megamind's conspiracy theory show inspired by *Coast to Coast AM,* and a night for alternative music.

"My favorite show is the heavy metal show," Ramirez said. It's called *Tales from the Pit,* and the group of prisoners who host it refer to themselves as "pit chiefs" and their listeners as the "pit crew." Lately, they've taken to referring to Ramirez as a pit chief, too, because he's written to them so often, he's become a part of the show.

In some ways, The Tank is like a community center for men who can never leave their cells. Aside from the music and the daily announcements, the DJs stream news and play soundtracks to movies. (The preferred genre is rom-coms, Hozaifeh confided—but "they really hate prison movies.")

There are also religious services, a Biblical rap show, suicide prevention programs and stock tips from death row. Sometimes, the men interview each other, and once they interviewed the warden. When I visited in October, they interviewed me.

I'd been so drawn in by Ramirez's enthusiasm during our conversation that I wanted to come back and see the station. The warden led me through a maze of walkways and hallways before we got to a tiny room buried inside the facility. From the outside, it looked like the door to a closet—but inside, the space was filled with sound equipment and computers. Except for the DJ's white prison uniform, the scene could have been inside an upstart studio anywhere in the outside world.

When Hozaifeh hit record, we talked some about my life—how I ended up in prison myself and how I became a reporter afterward. But I've been covering prisons in Texas long enough that a lot of the guys already know these things about me, and some sent in more idiosyncratic questions ahead of time: *What was your favorite thing on commissary? Do you like Madonna, Pearl Jam or Led Zeppelin? Pizza, steak or tofu?*

From their cells and bunks, the men of Polunsky steer the interview. It's an unlikely way to take some measure of control in the heavily regulated world of prison, and to hear their own words on the air at a station run by them and for them.

That's been part of the attraction for Jedidiah Murphy, who's been on death row for 20 years. Since he started listening to The Tank, he's been writing in to Megamind's conspiracy theory show regularly. Though the quirky content aligns with his interests, it's not the main attraction: It's the audience that doesn't judge him by his past, because they all have pasts, too.

"When you have people in prison that don't even really CARE about the crime or the situation, that is something that many of us have not seen," he wrote to me. "This is inmate-run for INMATES."

The guys running the radio station understand how much that means. They've never been on death row, but many of them—including Hozaifeh—have been in solitary, too, and they know how disorienting the constant isolation can become.

"You just don't know if you exist anymore," Hozaifeh said. "It just kind of removes your humanity from you, and I think the radio has put that back in the equation."

In September, a few days before Ramirez was to be executed for the 2004 killing of a store clerk, the guys who run the heavy metal show curated a playlist for him and played pre-recorded messages from his inside friends and outside supporters. The rap show read letters from listeners, recounting ways in which his contributions to the station had touched their lives.

As per usual, he tuned in—but this time he got to respond with his own voice. The day before Ramirez was scheduled to go to the death house, the warden made an unprecedented decision: He let the condemned man go to church. It was a special service outside, and there was a chain-link fence between Ramirez and the choir from General Population—"GP"—but it was still a first for death row. Afterward, The Tank aired the best bits for the whole prison to hear.

When Ramirez spoke, he talked about his regrets and described how he cried as he watched his mother walk away from her final visit. But he also talked about the radio station, and how it had given him one last chance to be part of a community.

"I don't know if y'all really understand how big that is because y'all in GP," he told the other prisoners. "Look at how y'all all next to each other. Y'all posted up, y'all walking around, y'all touching each other. We ain't got none of that. Y'all got community. We alone, we all by ourselves."

Before long, he'd be going somewhere else alone, taking the last steps to his own death in a sterile room an hour away in Huntsville. "Do you know how big that is?" he asked. "From all that I took out of the world, all the negative I did, all the people I hurt . . . all that selfish carelessness that I did as an idiot little kid, now I got to pay for it as a man."

As he talked, the men listening fell silent.

"For years now, the only thing I could do was make it about everyone else," Ramirez continued, explaining how he poured himself into the station in the hope that he could leave behind something good to help other people.

"Because it's important to me, man, it's important to me and that's all I can do. I'm alone. I'm alone in that cell. That's all I can do is give you my words."

One day later—on the night he was to be put to death—the U.S. Supreme Court decided to hear his appeal, halting the execution. Now, while he waits for the justices to weigh in, Ramirez is back on the row and tuning in to The Tank again, mailing Megamind his thoughts and contributions.*

When I left from my October visit to the station, I headed off in the opposite direction from which I'd come, thinking of Ramirez

* In early 2022, the Supreme Court ruled that Ramirez could have his pastor lay hands on him and pray as he was executed. Weeks later, the Nueces County District Attorney's Office asked for an execution date, then unsuccessfully tried to have it called off. Ramirez was put to death in October 2022.

and Hozaifeh and the little room filled with sound equipment. I flipped on my radio to 106.5 FM, and listened as Megamind pumped up his listeners, talking about faith and gratitude and finding ways to make meaning out of life behind bars. Just after I passed the Dollar General, his voice began to fade, replaced by the staticky words of a distant love song.

ORIGINALLY PUBLISHED BY THE MARSHALL PROJECT, IN PARTNERSHIP WITH THE *GUARDIAN*, DECEMBER 2021

TO THE SON OF THE VICTIM

BY SOPHIE HAIGNEY

I met you the day your father was shot and killed. I'd been in Oakland for a pink sunrise, watching police sweep a homeless encampment, gathering what we called "string" from residents who had nowhere—yet again—to go. I felt more outraged than usual and also maybe more useful. This was journalism, I suppose I was thinking, making sure the world knew what was happening right here. I wrote three hundred words for my newspaper's website in a café and was preparing to drive back across the Bay Bridge in brilliant golden morning light. Then I got a call.

An editor back at the office on Mission Street was listening to the police scanner and heard something unusual going on near Santa Rosa, about sixty miles northeast. Since I was already out, could I go? I could. I drove north, generalized dread already flushing cold through my veins, though I had no sense of what I was going toward. This is what the days were like, back then: waiting for something to happen, hoping it wouldn't, getting the call, driving, always driving, toward disaster.

There were black SWAT helicopters flying overhead and mixed reports from my editors: a robbery at two different addresses in Santa Rosa, three dead. Or maybe only one person was dead? Maybe they were related; maybe they weren't. Someone seemed to think that it had something to do with marijuana. I kept driving north into the brilliant sunlight, in a direction that—on other days or for other people—might have led to wine country or skiing in Tahoe.

Then the road turned into a vast sprawl of neon signs. I stopped at a gas station to buy a water bottle and a phone charger, a little shaky from hunger. I was listening to a song on repeat: "*Heard you were rolling in the good times out West, went to the desert to find your destiny and place . . .*"

I'd moved to San Francisco just a few months before to become a "breaking-news reporter." The romance of breaking news was that you were just thrown out there, learning on your feet, somehow transforming into a real reporter in the process. I had wanted this badly, all of it: the crime scenes and fires, the early-morning wake-ups and late-night phone calls. But it turned out I hated showing up on people's doorsteps in the wake of disaster and death. One Friday, there had been reports of a hostage situation many miles north. While the details emerged online and over the radio, I did something unforgivable in the profession: I went to the bathroom, took deep breaths, and waited a few minutes until someone else was sent instead.

The first Santa Rosa address was a bust. Or, rather, it wasn't really an address at all—it described a long stretch of halfway highway between two traffic lights. There were a few houses and I knocked on their doors but, to my relief, got no answer. I drove on, down a road that cut through farmland, where the distance between mailboxes grew longer. There were horses and shocks of green, as though drought had never struck here. This was the kind of place where neighbors could be relied upon to say, I can't believe that something like this would happen here of all places. Then I saw the Sonoma County Sheriff's truck hulking beside one of the mailboxes. This must be the place.

There were large cactuses and there was yellow tape. Even as I flashed my press pass, it was clear I wasn't going to get very close to the scene—to your house, a low, white ranch house I could see from the driveway behind the kind of padlocked gate that would keep horses in. A grizzled sheriff's deputy with a red beard and sunglasses looked at me with disgust. "The family's not interested in talking, ma'am," he said.

"Can you tell me what's going on?" I asked. Reporting under conditions like this was always full of roadblocks, and the primary obstacle was usually someone in uniform.

"You've got to call the press line," he said. He was disgusted by me, and I by him. Sometimes the only thing that motivated me in my reporting was the stoic "No" of police officers and sheriff's deputies and flacks on the phone. I pulled up nearby to wait. I fiddled with my phone, checked for new statements from the different law enforcement agencies, texted friends in New York—a boy I loved was there—and then looked up to see you, leaning on a gate and looking straight at me through the windshield.

I grabbed my notebook and scrambled to get out of the car, up to the gate. We stood for a minute in the dusty early-afternoon heat and didn't say anything.

You were about my age, give or take. I was twenty-two. You had been crying, though you made a good attempt at hiding it, red rims around your eyes.

I can't remember what I started to say, maybe something like, *Hi, I'm a reporter, I know today must be a hard day, but I was wondering if you could tell me a little more about—*

"We're not talking to anybody right now," you said quietly, looking down. I looked down, too, and saw you were wearing dark leather cowboy boots.

"I'm really sorry to bother you, it's just that"—I was trying to figure out what to do with my hands, gesturing too much, probably—"we're hearing reports that someone was killed here last night and I wanted to know if you could tell me if that's true?"

To that, you said nothing; you looked at me and turned away, walking back toward the house. The cop was watching me from the car with the window rolled down. Maybe he shook his head, or maybe that's something I've imagined since then.

I drove away, back to the first set of addresses, and fielded calls from frustrated editors. Someone from the family—perhaps you?—had spoken earlier to the *Press Democrat* and confirmed that a man had been tied up, tortured, and shot dead in the middle of the night. The suspects were a group of men who had mistaken the property for a cannabis farming operation—or perhaps it actually was one? Could I go back and find out? I could.

When I got there, I stayed in the car, engine on, watching the clock and hoping you wouldn't return. But you did, this time flanked by two men—boys?—who looked about your age. Maybe cousins or brothers or just friends. You recognized me and you looked my way almost imploring, as if to say, "I already *told* you: I need some time." The three of you walked toward a parked truck. I wish I'd given up then, but I instead followed numbly.

"You need to leave right now," said one of the other boys. I liked him for that.

"Can I just give you my phone number, in case you'd want to talk later?" I asked.

"Okay," you said, surprising me. But as I tried to write my number down, my pen ran dry. We stood awkwardly facing each other as I tried to sketch it, us just standing there in the brutal heat, your red-rimmed eyes behind sunglasses. Finally, I was ready to quit; I even shrugged. But then you pulled a pen out of your pocket, and you let me use it.

I have thought so often of this day—of my cruelty and your pain, of how powerful I was and how powerless I felt, of the pen you lent me, right in the moment of my defeat. What you must have felt that day remains impenetrable to me, even as I know more of the story, or at least as I know the bits of it that were reported in the following days, before everyone looked away. Your father was shot ten times while you and your mother were bound to chairs, with duct tape in your mouths. Four men and one woman were eventually arrested. More than a year later, a murder trial was ordered. It has yet to happen. That's one version of the story, but the ripple effects of that night are, I am sure, a much longer and more complicated story than could ever be written. No one has even tried, least of all me.

That day in Santa Rosa, I imagined that you hated me, but I now suspect you didn't think much of me at all. Probably I was part of the collateral of your grief, the random details—chipped nail polish, oddly shaped clouds, the color of someone's hat—that one notices in the moments before or after catastrophe. If you remember me at all,

I imagine that it's like that: a girl standing improbably in the glare of sunlight and rising dust, borrowing your pen after hers runs dry, the sound of gunshot still fresh in your ears. Regardless, you were generous to me on a day when you had no reason to be. I wish I'd been kinder in return.

ORIGINALLY PUBLISHED IN *LETTER TO A STRANGER: ESSAYS TO THE ONES WHO HAUNT US*, EDITED BY COLLEEN KINDER, AND REPUBLISHED IN THE *PARIS REVIEW*, MARCH 2022

THREE BODIES IN TEXAS

BY MALLIKA RAO

There's an optical illusion that went viral a few years ago, an illustration from a nineteenth-century German humor magazine. From one angle, the drawing looks like a duck; from another, a bunny. Over the years, many have weighed in on how our interpretation of the bunny-duck's ambiguity has to do with how we perceive and interpret the world. One study, by neuroscientist Peter Brugger, suggested that people in Switzerland saw a bunny more often in the springtime than in the fall. The British psychologist Richard Wiseman has found that high levels of creativity track with an ability to toggle easily between seeing both animals.

Ludwig Wittgenstein, before them both, discerned in the image a key to unravel the mystery of perception. Out of ambiguous raw matter a viewer suddenly perceives a duck, or a bunny. The moment of conversion and the resultant divergences fascinated Wittgenstein. What makes us perceive? What influences perception? How objective can any perception be?

I think of that optical illusion when I think of the case of Pallavi Dhawan. Asked to interpret an ambiguous scene, police officers in Frisco, Texas, made a decision: Pallavi was guilty. She had killed her ten-year-old son, likely by drowning him in a tub, where his body lay to rot. They arrested her. What details they found supported their narrative, one that maintained a presumption of guilt. Of course, it's possible they were able to see only one shape and couldn't consider another.

—

FRISCO ITSELF IS A SITE OF AMBIGUITY. THE CITY'S TWO LARGEST ETHNIC blocs are white, then Indian. One resident, interviewed by *The Dallas Morning News,* bought a home there sight unseen while she lived in India. The city is a version of India, if in ethereal, psychographic form, an overlaid map of beliefs and longings, of inheritances and expectations. Drawn sharpest perhaps where all the Indians live, the neighborhoods of Richwoods and Centennial, to the south, and the new, baby-treed developments off Coit Road and Independence Parkway, close to the schools where their kids show up every day, and to the enormous Karya Siddhi Hanuman Temple, studded with turrets and supersize stone elephants, an homage to Hinduism that rises out of the flat ground into the open Texas sky.

On January 25, 2014, Pallavi Dhawan woke, according to her

account, to find her son, Arnav, stiff and unresponsive beside her. The day before had been typical. She'd picked him up from Isbell Elementary, where he was in the fifth grade. He had done so well on a spelling test that she let him choose a reward: see a movie or go to the toy store. He ate grapes and chose a movie. By the time they got to the theater, though, he had fallen asleep, and woke up complaining that he was cold and tired. Toys"R"Us it was. Back at home, Arnav watched cartoons and said he was too tired to change into his pj's. He woke up twice that night, Pallavi said, and wandered into the living room, complaining of a chill. Finally she agreed to sleep next to him.

In the morning, Pallavi thought her son might be pretending to be asleep, trying to avoid the day. She asked him to wake up. She touched his cold skin and refused to believe what seemed evident, hoping desperately that he was faking. She picked him up and realized that his pants were wet with urine. No flutter moved his eyes, no breath his body. She took him to the tub. When she saw how he slumped against it, she knew he was dead. Something inside her still hoped. Again she asked him to wake up, frantically this time. *"Wake up, Arnav, wake up!"* She checked his pulse, his heartbeat. She pressed on his chest, blew into his mouth. She felt as if she'd gone into shock.

She gave him a bath, a Hindu custom for the recently deceased. She dressed him in his favorite clothes. In the kitchen, she first tried slipping ice into Ziploc bags, but that took too long. She used plastic grocery bags instead. She filled and knotted them, then took them to the tub and laid them around her son's body.

Eleven years earlier, Pallavi's husband, Sumeet Dhawan, had flown to India, where his father's dead body lay on ice in his family home, preserved in wait for him, the eldest son. Pallavi was pregnant with Arnav at the time, in her third trimester, so she could not go. But she heard of the particulars.

Now Sumeet was once again in India, at the tail end of a two-week business trip. The same duties that demanded he release his father's soul now applied to his son. As relatives in India had once waited for him to arrive from America, his wife now waited for the inverse. But Frisco, despite its ambiguity, is ultimately different from India. It's not a place where you can put a body on ice in a bathroom for days without societal repercussions, where you can wait quietly for a man to fly many miles across oceans so a soul can be properly released.

***

DEATH RITUALS VARY ACROSS INDIAN COMMUNITIES. MODERN LIFE BEING what it is, people have to adjust, adapt, and interpret old customs. When my own mom died, also in Texas, my brother did not set fire to her funeral pyre, as he might have in some other era, in India. Instead, he and I pressed a button at the same time on the crematory machine. Before that, at the funeral home—my brother in a lungi and me in a sari—we took steps around holy embers in an aluminum pan, guided by a priest as we petitioned god to bless the soul in the body we were soon to cremate. I insisted that I play an equal role to my brother in all the proceedings, and he agreed that was only right. A sort of accommodation was made, for our modern, feminist selves: I walked with him, though a step behind. Around a small pan, not a proper fire.

Pallavi wanted to do her duty by her son's soul. If she had called me that morning, I would have told her that exceptions are sometimes necessary. Ideally, Sumeet would preside over the cremation rites, but life is not always ideal. Because Sumeet wasn't there, maybe a priest could act as a substitute. Some other man, some other person. Maybe even you, Pallavi. Even though women never lit the funeral pyres, alongside my brother, I pressed the button. Times have changed. The rituals are meant to release the soul in good form to Yamaloka, the kingdom of the god of death, Yama. Interpretations to do with the particulars of death conflict across the religion, but one general idea is that Yama presides over a sort of courtroom. After his judgment, the soul continues the reincarnation cycle: another life on earth, beset both by the karmic debts and gifts of its last life. Each cycle offers another chance to attain moksha, full virtuousness, which ultimately lets a soul exit the cycle and reach the final, transcendent realm.

Within this calculus, every soul we encounter is on a long journey. Every meeting between people is brief but loaded. After a death occurs, the soul must be fed, kept strong as it makes its invisible way toward Yamaloka, and judgment. A white sheet is placed on the body, or gross form, and the feet are turned to face the south, the direction of Yamaloka. A circle of cow dung is to be drawn near a basil tree. Rice is rolled into balls and offered to the heavens as sustenance for the soul. But over the years, even Yama and his messengers must have noticed improvisations as migration led to a diaspora, and India itself changed. Rice is easy to find anywhere—cow dung, it depends where you are.

To some extent we're all victims of a game of telephone when it comes to these rituals. We owe them to ancient texts. The earlier texts are gentle in their view of the afterlife, but then comes the *Garuda Purana,* an authoritative codifier set down in Sanskrit, likely in the first millennium BCE. Over the centuries, versions have sprouted more versions, all influenced by competing traditions. The rules for rites collected in the *Garuda Purana* may be widely known, but they're not necessarily precisely known. A seminal English translation, published in 1911, has a gothic, poetic, and post-biblical tone; it speaks of hellscapes and paints Yama as a potentially brooding, fearsome judge, for certain unlucky souls. Some consider it dangerous even to read the text except during funerals.

According to the 1911 translation, all manner of harm and blessing we enact with our bodies affects the god essence inside us. The body is a vessel for the soul, and the soul is an embodiment of Brahma. A soul released from a life of sin sees things differently from a soul that's lucky to have lived inside a body engaged in virtue. To the unlucky soul, Yama appears in a dreadful form; to the virtuous, he appears radiant. Parents, by this reading, are custodians of children's souls from the start of their relationship. Pallavi could have found peace in her efforts to help Arnav live as best as he could on earth—if his death rites were not perfect, at least she'd have done that. But she seemed to want to do what she saw as her best for him until all her duties were done.

—

PALLAVI RECITED PRAYERS OVER HER SON'S BODY. SHE PRESSED TISSUE INTO his nostrils, a substitute for the customary cotton. She placed his favorite toys around his body and read from his favorite books. She did not want to call anyone whose presence might lead to an authority cutting into Arnav's flesh and disturbing his soul before Sumeet's work was done. Nor did she call Sumeet. He'd inherited a family condition: a weak heart. She did not want to endanger his life. Besides, their marriage was already so fraught that they hadn't talked since he'd left.

As for Sumeet, his return was delayed. Four days after Arnav's death, he came home. After greeting him at the door, Pallavi told him she was going to pick up Arnav from his Kumon class, still nervous—she would later explain—about Sumeet's heart. She drove to a gas station and bought a three-hundred-dollar Visa gift card and then went to a hotel, where she asked whether she could place a call from a room. She left after thirty minutes.

Sumeet turned on the television and tried to kick back. He checked his email and found a days-old note from Isbell Elementary, inquiring about Arnav's absences. As darkness fell, he became more agitated, worrying about Pallavi and Arnav out in the inclement weather. He called Kumon and discovered the boy had not been there all week.

Sumeet and Pallavi had moved to Frisco a little over a year ago to be close to his brother and sister-in-law, who lived nearby. They'd begun their lives as a family in Madison, Wisconsin, before a brief move to India. Arnav had health issues and required constant care. He'd presented challenges immediately after birth. His medical records detailed persistent problems: an unusually small skull (microcephaly), developmental delays, a brain cyst. The idea was to live near family and find support for Arnav, but eventually the couple decided to move back to America: to Frisco, Texas, this time. That way they could get family support as well as the top-notch medical and educational services that America, not India, could provide. But since the move, the fissures in their marriage had deepened. Sumeet had been in touch with a mental health hotline about Pallavi's depression and paranoia, and his own fear that she might kill herself. He had made a practice of calling the authorities for help. He once phoned the police after Pallavi dumped Goldfish crackers on the floor when he called the house messy. With the stress of caring for Arnav, perhaps it was understandable that she was depressed. And there had been a break-in at their house that might explain her paranoia. She had begun to lock the interior doors. The hotline representatives said nothing could be done unless Pallavi explicitly voiced a desire to end her life. As for the cops, there had been no crimes as such to report; an outburst with Goldfish crackers didn't count.

But now Arnav was missing. Sumeet called the mental health hotline and told them his wife had become truly paranoid. After his first call to them, she seemed convinced he was going to institutionalize her. He was terrified it had all come to a head. After he hung up, he called 911.

Two cops showed up, Butler and Adams. A missing person report couldn't be filed because Pallavi hadn't yet been absent twenty-four hours, but it did seem strange that she and Arnav were out at night and that the child's whereabouts for the past week were unknown. Strange, too, that Sumeet couldn't contact his wife, because she'd supposedly gotten rid of her cell phone. Sumeet's description suggested a number of possibilities: that she had kidnapped Arnav, or that she was mentally unstable—or perhaps something worse.

While Sumeet was talking to the police, Pallavi returned. She told the officers to wait, and then took her husband aside and said, at a volume they couldn't hear but he could, "He is no more." From Sumeet's exclamations and Pallavi's gestures, the cops pieced together that there was something behind a door. Based on this inference, Adams says he asked Pallavi, "Did you murder him?" And that, in response, she nodded.

They forced open the locked bathroom door and smelled the decomposing body before they saw it: a corpse in the tub. Adams thought he saw bruises on the skin, but no. What he saw was decay. One of the officers clamped handcuffs on Pallavi and took her to their car. Only then did they realize their body mics and the car cam—all standard recording devices—had somehow not been activated. They turned them on and asked Pallavi, now on record, "Did you murder him?" That is when she said—and on this point there is consensus—"You wouldn't understand."

—

THE USUAL MEDICAL EXAMINER AT THE MORGUE WAS OUT, AND IN HIS PLACE was Lynn Salzberger, red-haired, brisk-voiced, and an avid gardener. Salzberger had gotten into forensics because of its complexity. You are solving a puzzle using clues. It's a unique branch of science that merges a doctor's anatomical knowledge with the meticulous work of an investigator.

Arnav's body had arrived the night before. A damp, spoiled smell hung in the office that morning. Normally, Salzberger would talk to the mother in the event of a child's mysterious death, but Pallavi was in a jail cell, immediately treated as a suspect.

Salzberger noted the child's athletic pants, blue shirt, black briefs. What looked like rolled tissue in the right nostril. Well-formed limbs, still-strong teeth. She worked with the urgency demanded by a disintegrating, foul-smelling body.

No marks. Then again, the body was so decomposed, she wouldn't have expected to see much in that regard, anyway. The internal organs were in place, no abnormal collections of fluid. No clear internal injury. The mystery remained of why this ten-year-old boy would have suddenly died. Without visible injury or a confession, the strongest argument for homicide would be poison in the system.

A day or so later, Salzberger was at home drinking coffee. She paged open the morning paper and a story caught her eye: the boy in the tub. Arnav, she read, had had health problems, even an extensive file at the Mayo Clinic. Clues that she should have had as she faced his tricky corpse. Had she known, she might have worked differently, saved certain organs to send out to specialists.

By March, Salzberger had received Arnav's medical records and released her official autopsy report. The toxicology assessment had come back clean. She pronounced the death likely from natural yet undetermined causes. She was inclined to grant Pallavi the benefit of the doubt. Myocarditis, perhaps—inflammation of the heart. Or obstructive hydrocephalus, due to a brain cyst documented in Arnav's medical records.

She included a caveat: Unnatural causes could not be entirely ruled out. The mother had hardly acted according to the informal protocol of a grieving mother: no calls to family, friends, or authorities for help. Then again, the cops hadn't followed protocol, either. A list in the autopsy report titled, in part, "suspicious circumstances" begins with the absent medical records. Ultimately, "the autopsy didn't help elucidate anything," Salzberger told me. And so the report it produced mystifies. Look at it one way, and it says one thing. Another, and it says something else.

—

PALLAVI'S FRIENDS BACK IN MADISON, WISCONSIN, WHERE ARNAV WAS BORN, were shocked to see news photos of the drab, heavy woman in the orange prison jumpsuit. They remembered a cheery dynamo who had held down a demanding job as a software engineer, a manager of a team, someone who was adept at navigating an American corporate environment and happy to be in her new country, but who brought Indian mithai to share with classmates and the other moms on Hindu holidays as well. A woman who had it together, who could balance a lot. Funny and capable and up-front despite her challenges. The one who picked Arnav up from school, who raised him tirelessly while Sumeet was away on frequent work trips. She would never in a million years have hurt her son. She had quit her job a few years after he was born to care for him full-time.

Kalpana Kanwar, a Madison friend, met Pallavi at the Preschool for the Arts, a desirable place for the city's academics and professionals to send their kids. The two became friendly, linked by circumstance: Kalpana's son Dheer had autism, and both women were from India. During playdates, Pallavi taught Kalpana tricks—how to sneak bites of hot dog into Dheer's poha to up the protein content. How to sweep dust to the side of a room over the course of a week and then collect it all at the end.

One day, on a walk, Pallavi told Kalpana how she'd gone into her boss's office and said she needed a day off to buy proper clothes. She no longer had any time for herself. In her boss's office, Pallavi told Kalpana, she had started to cry.

Kalpana could see the enormity and compounding nature of her stress. She saw how demanding Arnav's needs were. Pallavi often took Arnav to hospital visits on her own. More and more often, Sumeet was away on business. Rumors of infidelity had started among close friends.

"Arnav may only live for fifteen years," Pallavi told Kalpana. This prediction is not reflected in the available records, but microcephaly is known to significantly shorten one's life span in many cases. Arnav was also prone to banging his head on hard surfaces. A family friend's wife wrote in an affidavit that the pressure in the boy's skull caused intense pain. Arnav took his anger out on Pallavi mainly, but also on other kids. He frequently wouldn't eat. It sometimes seemed like if she wasn't with him every moment, he wouldn't survive.

To Kalpana, it was clear that Pallavi was the parent who was in touch with what Arnav needed. When the family moved to India in 2008—first to Hyderabad, then to Delhi, hoping for support from relatives—the divide between Pallavi and Sumeet only grew. They had known each other as friends before marriage, but theirs was an arranged alliance; it came about through the orchestration and consent of family members. As a result, Pallavi could seem at the mercy of her in-laws' judgment, and her husband could seem to hold loyalties to his old family rather than their new one. On one occasion, Pallavi told Kalpana, Sumeet neglected to invite her out to dinner with his friends and their wives. He didn't concede until his friends said he must. From what Kalpana could tell, her challenges were recognized neither by her in-laws nor by her husband, and India offered little institutional or societal understanding of special needs. As a result, Pallavi seemed to be generally misunderstood and underappreciated, became a scapegoat for all manner of problems, and was presumed to have invented them. One of Sumeet's friends in India would later describe how a group of them figured she was a typical American parent, overprotective and oversensitive.

"There's a book for little kids about filling your bucket with marbles," Kalpana told me. "I felt her life—continuing in Texas when they moved there—was just a continuous process of her bucket getting emptied. When I read about what happened . . . To me she was a woman whose bucket was pretty much empty at that point."

—

MAYBE THERE'S NO OPTICAL ILLUSION QUITE LIKE A MARRIAGE. THERE'S THE image of it presented to outsiders, but glimpses of its inner reality can complicate that image. There are the perspectives held by each partner: one person's version of reality against another's, the two often mystifyingly at odds. How can both be the one who's always doing the dishes?

The task fell to Sumeet to cohere worlds: their family's private one with the public one. He felt hysterical and confused by what was happening. He had said who-knew-what to the police officers, leaning on his old habit of casting blame at his wife, perhaps. Then she had vanished in what felt like seconds, carted away. He still had sweets and toys in his suitcase to give to Arnav, but his reality had changed so dramatically. It felt like ages before the paramedics arrived. They told him Arnav was dead. He tried over and over to go into the bathroom and heard only "No." They took away the body. He called his brother, who lived nearby. Everyone had questions, but he had no answers.

Hours had passed since the officers left and it was now deep, dark night. Sumeet was called into the police station. His brother went with him. They waited in a room and talked a bit, and then Sumeet met with a detective and shared the facts that seemed important: Arnav was a child with special needs, getting specialized services from the Frisco school district. He handed over documents he'd pulled in the chaos of the evening, school records from Madison, that mentioned Arnav's microcephaly and developmental delays. He said how devoted a mother Pallavi had been. He asked that the investigation unfold calmly.

He slept at his brother and sister-in-law's. The next morning, Sumeet saw the top news item on the Frisco PD website. Pallavi was being charged with murder, was suspected of drowning their son.

What was happening? Through friends, he tracked down a lawyer named David Finn.

Finn was en route to his downtown Dallas office from an outlying courthouse when he got the call. He agreed to detour to the Frisco jail, where he met Pallavi, outfitted in an orange jumpsuit and sitting behind glass. He sensed a good person who needed help—his newest client. He met Sumeet and they headed to the station to try to get the keys to the house, which the police had confiscated. While they waited to meet with a detective, Sumeet received a call. It was the medical examiner's office. Natural causes were still "very much" on the table. So why the arrest? The men were interrupted by an onrush of press members. A reporter, Shaun Rabb, informed them that a press conference was taking place. Pallavi was being charged, it seemed, on the strength of a confession based on a nod.

At the Dhawans' house, on Mountain View Lane, Finn and Sumeet talked at length. Finn started to understand the pieces of a story the cops seemed to have missed: a break-in at their house; medical records and other documents locked for safekeeping in the car trunk. Sumeet told Finn that he had tried in vain to explain to authorities about the records and their confusing placement in the trunk but hadn't found success. A few days later Finn drove to the courthouse to view the warrant on the car. He saw words validating Sumeet's claims listed in the official inventory: medical records, including those from the Mayo Clinic. Finn went on TV. *The Dallas Morning News* picked up the story—and that was how Lynn Salzberger came to know there was a medical history at all.

And yet Pallavi herself had told Butler and Adams about these records right after her arrest. Their newly activated recording devices had picked up her words about how the boy's medical records were in the trunk of the family car. But the police hadn't seemed to grasp that those records might be important. Or maybe they hadn't heard her.

—

ON A DAY IN EARLY FEBRUARY, ABOUT A WEEK AFTER THE ARREST, DETECTIVE Wade Hornsby called Sumeet on his cell phone. Pallavi had been out on bond for a few days, after the judge had reduced her bail from a six- to a five-figure number. Friends had hustled to get the money together after the immediate frenzy of the arrest. Now both the Dhawans and Finn were on Mountain View Lane, preparing for a vigil. Classmates of Arnav's, family members, neighbors, and community well-wishers were to arrive in three hours. Still, Hornsby wanted to talk to Sumeet. Could he come into the station? The vigil was soon to begin, but Hornsby, Finn says, was adamant.

At the station, the detective mentioned a conflict of interest. Finn was Pallavi's lawyer, not Sumeet's. That conflict, Finn said, could be waived, as Hornsby also knew, and on the spot, Sumeet agreed. Finn became his lawyer, too. As they walked to a back room, Hornsby turned and looked over his shoulder, according to Finn, and asked, "So, Finn: You giving them some sort of frequent-flyer family discount?"

Finn will tell you he doesn't like bullies. He's got a kid with epilepsy and he's Irish Catholic. He figures it's in his blood. The fighting spirit. "Down with the Brits," with all oppressors, really. At his church in Dallas, he helps operate an aid program for refugees. Finn and his fellow congregants act as chauffeurs, donate clothes, get to know kids and parents from Sudan, from Burma. He became a defense attorney, after all, a guard dog, after having served as both a prosecutor and a judge. His ex-wife used to say he "leans into the punch."

And he didn't like Hornsby's tone. His clients, he could see, were gentle. He'd felt a bond forming with Pallavi from the moment he met her at the jailhouse—he knew what it was like to care for a child with special needs. And Sumeet was like a scared rabbit, a man who seemed to feel he was not entitled to respect. This station was like a Texas schoolyard: they were the foreign targets and Hornsby—he was a bully.

In a back room, Hornsby got meaner, by Finn's account. Sneering and disbelieving, adamant that Sumeet had never mentioned any special needs—"That's baloney," Hornsby supposedly said. Was this any way to treat a man whose son was dead, who was only trying to help the investigation by illuminating a crucial aspect the cops had missed—one they now seemed to feel the need to erase from his family's history?

Finn ended the meeting. There was an interview with Pallavi scheduled to air on TV that night, the first footage of her the public would see, an exclusive with Shaun Rabb, the reporter Finn had spoken with at the station. "Hey, Wade," Finn asked before he left, using the detective's first name expressly to bother him. "You like your job?"

Is that a threat?

"I'm just saying. You might want to watch the five o'clock news tonight. You're about to get lit up."

—

MAYBE IT MADE SENSE THAT FINN WAS "TRYING THIS CASE IN THE PRESS," AS the Frisco mayor put it. The reading public in the rapidly changing state seemed interested in ambiguity and tired of the old guard. The Indian head nod—an emblem of ambiguity; a nod that can famously mean yes or no—became a favorite point of discussion. If an arrest based on a nod was already suspect, an arrest of an Indian woman based on a nod was doubly so, went a line of defense built in online articles and Facebook commentary, linked to a judgment about Texas cops: their once-excusable parochialism was at this point a menace to the public.

But Pallavi said she hadn't nodded at all. In their affidavits, she and Sumeet wrote that the entire exchange with the cops—the question, the nod in response—hadn't actually happened. She wasn't a murderer who nodded, nor was she a saintly, hapless immigrant who bobbed her head.

In August, an examining trial revealed the tensions of this caricatured state of affairs up close. Hornsby described a case built on the observations of people whose authority went unchallenged: the officers on whose word alone the nod had taken place. Teachers who, he said, described Arnav as perfectly normal. Hornsby even googled "arachnoid cyst" as soon as he heard of the ignored medical records to see for himself the seriousness of Arnav's condition, a gesture Finn mocked, as if it betrayed a wishfulness and desperation those searches can seem to suggest. Maybe Google could tell Hornsby what he wanted, could validate this oversight by the cops. Meanwhile, any insights from Sumeet and Pallavi did not seem to have been heard, much less sought out—Hornsby, in cross-examination, said that Sumeet had told him about the precedent set by his father-in-law's body only "later"; and with regard to the snafu with the medical examiner and the records, "We were not aware, so we did not share," he said.

On the other side, Finn's peacocking could seem destructive to his clients' case. "I have lived in India," he told the court, in reference to a time after college when he'd backpacked through the north of the country. Had Hornsby ever visited? The judge deemed the matter irrelevant. On the subject of funeral homes in India, Finn goaded Hornsby, "Did you google that, too?" Yet, Finn then went in the other direction, with an assertion that not a single such home exists in India. Finn's mockery of the Frisco police force's seeming unworldliness in contrast with his own sophistication can seem, in hindsight, to flatten all parties—including himself and his clients—especially given the relative shallowness of his own authority on India, revealing a larger, blustery underestimation of the complexity of a place far away, and of the two people in his charge. In this environment, Pallavi received her first hearing. The judge declared there was reason enough to keep the case going. A grand jury trial was set for September.

It never came to pass. On September 4, 2014, not quite nine months after Arnav's death, Pallavi and Sumeet Dhawan were both found dead at their home. They were determined to have ingested sleeping pills—Pallavi, a fatal dose; Sumeet, not enough to kill him, according to an autopsy report. He'd died from a fatal blow to the head by a cricket bat. Pallavi had drowned in the backyard swimming pool. The couple were days away from the grand jury trial that would either convict or acquit her.

—

ON FACEBOOK, A GROUP CALLED THE CORRUPT FRISCO TEXAS COPS DOCUMENTS the police force's supposed misdeeds, large and small. A wrongful arrest of a Black woman holding a sign. A bribe taken. The page is run by an anonymous white man who says he was "radicalized" after a faulty arrest at his home in Frisco. When he saw the news about the Dhawans' deaths, he cried. He assumed it was a double suicide, he told me. He saw another instance of police corruption—in this case, a grueling slow play of an investigation around an arrest that never should have happened.

By then the confusing, drawn-out duel between the police department and Finn could be traced in the local papers, with the couple at its center. A final spat had concerned the police's refusal to return the Dhawans' impounded car, a loaned Lexus, unless the couple agreed not to sue for damages. The cops intended to force a confession as well as to exonerate themselves, Finn insisted.

But the public trial he was conducting never reached a satisfying end. The media interest died in part when the couple did, with a final plea from Finn: A note had been found near the bodies of Pallavi and Sumeet. Its contents still have not been made public, despite Finn's filing of multiple requests. My own Freedom of Information Act requests to the State of Texas went unanswered.

Finn has his theories on what the note might say. Before their deaths, the couple had requested permission from the judge to travel to India for a ritual ceremony for Arnav—more duties. Their request had been denied. Finn sees this as the "last straw" for Pallavi. To his mind, the note likely blamed the Frisco Police Department, and Sumeet for his own misdeeds, and perhaps included a little shot of sunlight for Finn: "If I were a betting man, I'd bet she apologized for letting me down. They wanted a confession from her and they weren't going to get it. They jumped the gun, wound up looking silly, stupid, racist, incompetent. And, unfortunately . . . Pallavi gave them the only way out."

The suggestion of murder in effect recast Pallavi. "It just kind of makes you scratch your head and wonder," Salzberger told me. "Well, if she killed the husband, and she's mentally unstable, then she easily could have killed the child, too."

When Salzberger told me that, I heard an overly clean sense of personhood that runs through so much of this case, a seeming resistance to grant Pallavi total humanity. Salzberger herself linked Sumeet's infidelities with his death. And several anonymous parties speculated to me that he'd been cheating during the very time when Arnav died, that the depth of that betrayal, her discovery of it, led to Pallavi's act. Would a woman in such a context necessarily be an unhinged one, as per Salzberger's formulation? Aren't vengeful women in the abstract often celebrated, as mascots, revealers of the ills of society, more sane in some ways than the rest of us? *Gone Girl, Thelma & Louise.* "You shoot off a guy's head with his pants down, believe me, Texas is not the place you want to get caught," Louise tells Thelma, heroines to the last, killers of a rapist and, in turn, truth tellers about the nature of this country, of womanhood. White women wronged by men who cheat in various ways are allowed a capacity for violence as emissaries of truth: They show us how hard it is to be a woman. They live in a moral gray space engendered by their birth. But Pallavi seemed not to take a rightful place even in this mythology, seemed to exist in people's minds as either guilty or innocent, mad or saintly.

Those who knew Pallavi personally seem able to grant more psychological nuance to their friend. Geri Gibbons, a fellow mom from the Madison preschool, suggested both innocence and guilt. Guilt as a by-product of the toll the criminal justice system can take on someone whose humanity it does not keep in mind. It was the state who drove her friend to the edge. "Only in Texas," she told me, "can you take somebody, accuse them of a murder they didn't do, and turn them into a murderer."

To me, blame feels impossible to place. The tragedy, in its sheer finality, feels almost inevitable, set in motion by unstoppable forces:

Shakespearean. Three people—an entire family—wiped out. From one angle, the involvement of the police can be seen as unrelated, as if the cops stumbled into something they were never meant to see.

—

IN DECEMBER 2020, I DROVE TO THE KARYA SIDDHI HANUMAN TEMPLE. I TOOK photos of the intricately carved stone against the blue sky: those elephants, those turrets. A friend from Delhi to whom I sent a photo said it looked like India. I felt that, too, sitting in the parking lot.

I drove twenty minutes to Isbell Elementary. Walking around the campus, I felt like an interloper, haunted by memories that didn't belong to me. I imagined Arnav walking into the school, through the glass doors. In Madison, Pallavi and Arnav had both made friends at his school. Yet I could find no one who seemed to consider them friends here.

One Isbell mother told me her son had made a drawing of Super Mario for the vigil at the Dhawan family home soon after Arnav's death. He'd felt sad and confused. But he and Arnav hadn't been friends, exactly. Arnav had chased him once during a class field trip, and she had had to chide him. Her son also had some physical ailments that made him delicate. Later she emailed me a Facebook post she had shared around the time of the vigil; the one-year anniversary of Arnav's death had prompted it to resurface. Back in 2012, she had posted a news story and wrote of how proud she was of her son for managing the troubling experience of the death of a friend with grace, for having the bravery to talk to the media. Reading her words, I saw clear maternal pride, for her own son—even in the midst of a story of another mother's loss, of another boy's destruction. I saw her love and compassion, and I also felt lonely taking in how she framed that love to her followers. Pallavi and Arnav were mentioned nowhere in her words. More generally, the Dhawans' story seemed apt material onto which people projected all sorts of self-directed narratives.

I drove to the house on Mountain View Lane. It looked, to my eyes, at once totally ordinary and ghoulish. On display in the front yard was an elaborate Christmas manger scene, featuring life-size statues of Jesus, Mary, and Joseph. In the back drive I spotted a parked speedboat. I was almost certain this house wasn't owned by Indians anymore.

Why did I care about a woman I'd never met, who sometimes scared me when I read the clinical notes about her life? Perhaps I was also projecting. I recognized something in her, maybe, based in part on our shared history. My parents, Indian immigrants, moved in the early 1980s to Texas, where I was born and raised, before I ultimately set off to a life in New York that felt less encumbered by regional racism. I suppose this projection felt worthwhile, though, tied up in a kind of instinctive empathy that she was cheated out of in life. I thought of the only public footage of Pallavi, shot for the evening news a week after her arrest. She sits with Shaun Rabb, with David Finn to her left and Sumeet to her right. She speaks slowly and methodically. Her voice starts to break when she describes touching Arnav's body, taking him to the tub. She was certain she could revive him with water. Her eyes are full of pain. She demonstrates with her hands how he slumped against the tub when she placed him inside. She wipes away tears. The movements feel authentic. "My only thought right now, Shaun," she says, "at the time also, and even when they took me in: I felt I did my part for waiting for his father, and from there on it's up to his father. I needed to do that for my kid. That was important for me as a mother, for that child."

In time, I would read of two arrests that happened just before

Pallavi's, of Purvi Patel, and then of Bei Bei Shuai, together the first women ever to be tried for feticide in America. "It's no coincidence that both of these women are of Asian descent," the activist Miriam Yeung wrote of those seminal cases in a 2015 *Washington Post* op-ed. "Asian women have been singled out when it comes to criminalized reproduction because of ugly stereotypes that claim we have a disregard for life." An officer interrogating Patel in the hospital asked her repeatedly of the fetus's father, "Was he Indian, too?" Later, lost in my research, I'd read of how British missionaries in India in the eighteenth century coined their own term for Hindu mothers who seemed not to care about the preciousness of life, who seemed to see the value of life differently: *unnatural mothers*.

Yet I'd seen the opposite. Sitting at my desk in a newsroom many miles away, I'd watched the TV footage of Pallavi and I'd thought of my own mother, who might have adhered to religious laws over the laws of the land, who might have seen her duty to me in pragmatic terms that would be mystifying to outsiders. Arnav's soul had had a short go of it in his life on earth—it was Pallavi's duty to ensure that his soul had a next life. I understood the logic underpinning what might sound like a fanciful, outrageous thought. I remembered how our own friends had arrived during the eleven days of mourning after my mother's death, scientists and other professionals. A priest who made movements in the corners of our sunroom to shoo the soul from the house. I remembered closing my eyes and thinking intensely of her soul—an entity that felt to me, in that moment, childlike, innocent. I remembered how I told this being with all the silent force I could muster that we didn't need or want her to linger, that her passage to the next life was waiting. We would be fine. I would be fine. Her duty to us was done.

In all the media accounts of Pallavi, I didn't see her treated as an ordinary human being with a complex and accessible sense of logic—but to me she felt that way. I had no doubt this was because of the logic system I'd understood by virtue of being born to Hindus in Texas, one of whom died there. I understood warring systems, for that matter, a Hindu sense of order, a displaced sense, Texan, American. All around me were stories trumpeting the power of representation—in the movies, on TV—but I wondered what it might be like for a Texan cop to face a woman born in India, still early in the generations-long process of assimilation, who had turned inward and had become impenetrable due to her hard family life. Would he know what to make of her? Would her behavior merely affirm his presumptions? Finn told me he had a body-language expert scan the TV footage, and this expert had been set to testify at the grand jury trial that Pallavi had all the markings of a truth teller, a woman in grief. I didn't need an expert to tell me that. I felt it just watching her.

Finn hypothesized to me an alternate universe theory of a white woman. Had the cops found a white mother in a similarly fraught setting with a dead child, they'd have treated her as a grieving woman first and a potential suspect second, he mused. In this case in particular, that distinction would have made a difference. So much seemed to have been set in motion by Pallavi's arrest, which barred her from offering insight as the boy's mother to those in charge of the investigation; that arrest, moreover, had been built on a first impression, made perhaps in error, that could neither be acknowledged nor even properly understood.

Even among those guilty of killing their children, a divide can seem to separate women across races. In my research, I saw that white women who face trials in America after the death of their children are often spared, if at all, by a ruling of not guilty on reasons of insanity. The phenomenon hearkens back to an old legal distinction, a defunct network of laws stretching from Europe to America, based on the debunked concept of "lactational insanity," which suggested that the hormonal shifts in a new mother could explain derangement. Although ultimately discarded from formal use, that line of thought can nonetheless seem to reflect a certain leniency in the allowance of grace to some women over others. Even as the missionaries of Britain were pronouncing certain Indian mothers to be "unnatural," the legal minds of the nation were at work articulating a reason behind those unnatural behaviors when they took place among white bodies on English soil.

Another organizing structure hems Pallavi in as well, to my ears. I heard in her words, in that sole snatch of footage, a coded message to Sumeet. Given what I know, she sounds almost as if she is challenging him to do his duty. Sumeet, in his affidavit, blames himself for her arrest, for the confusion. He cites his "male ego" as the source of his inability to understand the constant stress on Pallavi. He questions whether he actually said all the things the cops suggest he did—but no matter, he writes: he is not to be trusted, especially not in the frenzied moments after discovering his son's death. When his own father died, he writes, he cast blame everywhere. Here, too, he unfairly shone a spotlight on his wife—or seemed to, before understanding his error.

Lynn Salzberger shared with me a revenge paradigm sometimes associated with cases like this. One spouse is cheating and the other harms the child to cause their spouse pain. Salzberger meant to explain a version of the story where Pallavi did kill Arnav, but I heard another possibility. Even if Arnav died naturally, it seemed possible to me that Pallavi could have been exacting a form of revenge on Sumeet afterward, one that started with insisting he do his duty and perhaps ended with his murder. Kalpana Kanwar told me she wouldn't be surprised if Sumeet's family had been encouraging him to leave her, to start a new family, a theory that Hornsby also proposed during the sentencing trial, based on a translated snippet of the conversation between Sumeet and his brother at the station, which had been recorded, unbeknownst to them. I could understand how a woman in an arranged marriage might feel a bit like a pawn, a hired employee meant to produce a certain type of child and family. When the building of a family is a group affair, a woman can become an outsider. If her son dies and her half-hearted, unfaithful husband is away, she might want to make extra sure he finally feels the weight of his responsibilities. *From there on, it's up to his father.*

That first night at the station, Sumeet and his brother had spoken in Hindi, apparently discussing how he should have left Pallavi sooner, how his latest trip to India had been made to set that move in place. The conversation suggested a partial explanation for the bullishness of the Frisco PD in sticking to their charge: they were leaning on Salzberger's revenge paradigm to push a murder theory. But my eye latched on to something else as I read this account in the court transcript, something that irked me as I was meant to take in this Frisco PD narrative. The word *Hindu* appears multiple times in the trial transcript, when the correct word is *Hindi*. As I saw that repeated typo, I wondered if the error had been the court reporter's or if it had actually been spoken by those in charge of Pallavi's fate, in that courtroom. I wondered how much of any case is built and tried on fact and how much on feeling, on instinct. No one in the court had been of Indian origin except the defendant and her husband. This was a part of Frisco not built for them.

Not that anyone I know who came to America from India wants it to be, exactly. My own family members who arrived in Texas after we did settled in Frisco, Coppell, cities at the edge of the Dallas metroplex that could take them, even if that had never been the city's plan. One cousin told me she remembered something particular about this case: how impossible it had been even for Indian people to understand it. She told me how a relative in Coppell had puzzled over how an Indian woman of Pallavi's pedigree could have been embroiled in such a sordid affair. These sorts of dramas are not for certain immigrants. Not for our kind, either, the inclination to enter the police force from the other, proper side, to take on jobs embedded in the grittier corridors of a society, where so much that is internal is made to function. We don't commit crimes or, on the other hand, raise our kids to become cops. An Indian body was never meant to play any role in this optical illusion.

I drove to Isbell Elementary one last time on my way back to the house I grew up in, where we had thrown pujas and birthday parties, where we had lived feeling welcome and not, where one Thanksgiving, two cousins wearing lungis as they changed the oil on my dad's car in the driveway were questioned by cops responding to a call from a neighbor. And we laughed afterward at the thought of the officers seeing our front door open to another brown man, my father, also in a lungi. I wondered if the Dhawans had walked to the small park next to the school, the way families were doing this day. I wondered if Arnav had run on the winter grass, drained of color, so like the grass I'd run on as a kid, in some part of Texas that was secretly two things at once.

ORIGINALLY PUBLISHED IN THE *BELIEVER*, MARCH 2022

ACKNOWLEDGMENTS

Evidence of Things Seen is both a companion volume to and an extension of my previous anthology, *Unspeakable Acts.* It continues to surprise and delight me that the appetite for true crime storytelling that provokes, interrogates, and challenges long-standing assumptions remains strong. That continued strength owes to a great many people.

Deepest thanks to my agents, David Patterson and Aemilia Phillips, as well as Chandler Wickers, Hannah Schwartz, and the entire SKLA faithful team; to my editor, Sara Birmingham, who took on and shepherded this project with gusto and enthusiasm; to my wonderful and long-standing publisher, Ecco (four books in and still going!), and especially TJ Calhoun, Miriam Parker, Meghan Deans, Sonya Cheuse, and Helen Atsma; to Allison Saltzman for the outstanding cover; to the HarperCollins sales force and library marketing team, with perpetual gratitude for their hard work on my books' behalf; to the booksellers, particularly the independent stores, who continue to share in my passion and excitement for crime stories, especially those that break the mold; to Rabia

Chaudry for her thought-provoking introduction and perennial inspiration; to all of the talented and dynamic contributors included in this anthology; and to family, friends, and fellow writers who provided untold support as I put together *Evidence of Things Seen*.

OTHER NOTABLE CRIME STORIES

WHAT TO READ, LISTEN TO, AND WATCH

ESSAYS AND FEATURES

- "True-Crime Fanatics on the Hunt: Inside the World of Amateur Detectives" by Ellie Abraham (*Guardian,* March 2021)
- "A New Face of American Gun Ownership" by Agya K. Aning (The Trace, February 2022)
- "'Nothing Will Be the Same': A Prison Town Weighs a Future Without a Prison" by Tim Arango (*New York Times,* January 2022)
- "Inside Trump and Barr's Last-Minute Killing Spree" by Isaac Arnsdorf (ProPublica, December 2020)
- "Have You Seen These 51 Women?" by Ben Austen (*Chicago Reader,* January 2021)
- "Punishment by Pandemic" by Rachel Aviv (*The New Yorker,* June 2020)
- "I Write About the Law. But Could I Really Help Free a Prisoner?" by Emily Bazelon (*New York Times Magazine,* June 2021)

- "True Crime Is Rotting Our Brains" by Emma Berquist (Gawker, October 2021)
- "Inside the Once-Controversial Trend That Took Over True Crime TV" by Meredith Blake (*Los Angeles Times,* October 2020)
- "Witnesses to the Execution" by Keri Blakinger and Maurice Chammah (The Marshall Project, July 2020)
- "The Man Without a Name" by Katya Cengel (Vox, August 2020)
- "When James Baldwin Wrote About the Atlanta Child Murders" by Casey Cep (*The New Yorker,* May 2020)
- "Incarcerated and Invisible" by Gray Chapman (*Atlanta,* March 2022)
- "Telling Stories About Crime Is Hard. That's No Excuse for Not Doing Better" by Jason Cherkis (*Columbia Journalism Review,* August 2020)
- "When the Misdiagnosis Is Child Abuse" by Stephanie Clifford (*Atlantic,* August 2020)
- "The Life Breonna Taylor Lived, in the Words of Her Mother" by Ta-Nehisi Coates (*Vanity Fair,* August 2020)
- "The Murders Down the Hall" by Greg Donahue (*New York Magazine,* October 2021)
- "Secrets of the Death Chamber" by Chiara Eisner (The State, November 2021)
- "How Crime Stoppers Hotlines Encourage Sketchy Tips and Hurt Poor Defendants" by Tana Ganeva (*New Republic,* October 2021)
- "Dubious Alternative Lyme Treatments Are Killing Patients" by Lindsay Gellman (*Bloomberg Businessweek,* October 2020)

- "Michelle Remembers" by Jen Gerson (*Capital Daily,* August 2020)
- "The Growing Criminalization of Pregnancy" by Melissa Gira Grant (*New Republic,* May 2022)
- "In Puerto Rico, an Epidemic of Domestic Violence Hides in Plain Sight" by Andrea Gonzalez-Ramirez (GEN/Medium, June 2020)
- "The Enduring, Pernicious Whiteness of True Crime" by Elon Green (The Appeal, August 2020)
- "The Shadow and the Ghost" by Christine Grimaldi (*Atavist Magazine,* May 2022)
- "They Agreed to Meet Their Mother's Killer. Then Tragedy Struck Again" by Eli Hager (The Marshall Project, July 2020)
- "Catching Predators with *Riverdale* Mom Marisol Nichols" by Erika Hayasaki (*Marie Claire,* May 2020)
- "The Ballad of the Chowchilla Bus Kidnapping" by Kaleb Horton (Vox, July 2021)
- "The Store That Called the Cops on George Floyd" by Aymann Ismail (*Slate,* October 2020)
- "Inside the Hunt for Christine Jessop's Real Killer" by Malcolm Johnston (*Toronto Life,* November 2021)
- "Murder in Old Barns" by Lindsay Jones (*The Walrus,* June 2020)
- "A Bestselling Author Became Obsessed with Freeing a Man from Prison. It Nearly Ruined Her Life" by Abbott Kahler (The Marshall Project/The Cut, March 2021)
- "Philadelphia Keeps Revisiting the MOVE Bombing Because We Never Truly Learned It" by Akela Lacy (*Philadelphia Inquirer,* June 2021)

- "Homicide at Rough Point" by Peter Lance (*Vanity Fair,* July 2020)
- "The Evidence Against Her" by Justine van der Leun (GEN/Medium, June 2020)
- "Targeted" by Kathleen McGrory and Neil Bedi (*Tampa Bay Times,* September 2020)
- "White Riot" by Laura Nahmias (*New York Magazine,* October 2021)
- "He Said, They Said: Inside the Trial of Matthew McKnight" by Jana G. Pruden (*Globe and Mail,* July 2020)
- "The Pretender" by Josh Rosengren (*Atavist Magazine,* October 2020)
- "Kip Kinkel Is Ready to Speak" by Jessica Schulberg (*HuffPost,* June 2021)
- "Inside eBay's Cockroach Cult" by David Streitfeld (*New York Times,* October 2020)
- "Two Wealthy Sri Lankan Brothers Became Suicide Bombers. But Why?" by Samanth Subramanian (*New York Times Magazine,* July 2020)
- "What We Still Don't Know About Emmett Till's Murder" by Wright Thompson (*Atlantic,* September 2021)
- "The Wildest Insurance Fraud Texas Has Ever Seen" by Katy Vine (*Texas Monthly,* August 2020)
- "The Little Cards That Tell Police 'Let's Forget This Ever Happened'" by Katie Way (*Vice,* September 2020)
- "One Roadblock to Police Reform: Veteran Officers Who Train Recruits" by Simone Weichselbaum (The Marshall Project, July 2020)

BOOKS: NARRATIVE NONFICTION

- Maurice Chammah, *Let the Lord Sort Them* (Crown, 2021)
- Becky Cooper, *We Keep the Dead Close* (Grand Central Publishing, 2020)
- Emma Copley Eisenberg, *The Third Rainbow Girl* (Hachette Books, 2020)
- Nicole Eustace, *Covered with Night* (Liveright, 2021)
- Sonia Faleiro, *The Good Girls: An Ordinary Killing* (Grove Press, 2021)
- Justin Fenton, *We Own This City* (Random House, 2021)
- Margalit Fox, *The Confidence Men* (Random House, 2021)
- Gus Garcia-Roberts, *Jimmy the King* (PublicAffairs, 2022)
- Elon Green, *Last Call* (Celadon Books, 2021)
- Elizabeth Greenwood, *Love Lockdown* (Simon & Schuster, 2020)
- Nicholas Griffin, *The Year of Dangerous Days* (37INK, 2020)
- Kathleen Hale, *Slenderman* (Grove, 2022)
- Evan Hughes, *The Hard Sell* (Doubleday, 2022)
- Dean Jobb, *The Case of the Murderous Dr. Cream* (Algonquin Books, 2021)
- Chris Joyner, *The Three Death Sentences of Clarence Henderson* (Abrams, 2022)
- Patrick Radden Keefe, *Empire of Pain* (Doubleday, 2021)
- Jarett Kobek, *Motor Spirit* (We Heard You Like Books, 2022)
- Kathryn Miles, *Trailed* (Algonquin Books, 2022)

- Sierra Crane Murdoch, *Yellow Bird* (Random House, 2021)
- Robert Samuels and Toluse Olorunnipa, *His Name Is George Floyd* (Viking, 2022)
- Ravi Somaiya, *The Golden Thread* (Twelve, 2020)
- Leah Sottile, *When the Moon Turns to Blood* (Twelve, 2022)
- Kate Summerscale, *The Haunting of Alma Fielding* (Penguin Press, 2021)
- Sarah Weinman, *Scoundrel* (Ecco, 2022)
- Phoebe Zerwick, *Beyond Innocence* (Atlantic Monthly Press, 2022)

BOOKS: MEMOIR, ESSAYS, CRITICISM

- Keri Blakinger, *Corrections in Ink* (St. Martin's Press, 2022)
- Hilary Fitzgerald Campbell, *Murder Book: A Graphic Novel of a True Crime Obsession* (Andrews McMeel, 2021)
- Katherine Dykstra, *What Happened to Paula* (W.W. Norton, 2021)
- M. Chris Fabricant, *Junk Science and the American Criminal Justice System* (Akashic Books, 2022)
- Debora Harding, *Dancing with the Octopus* (Bloomsbury, 2020)
- Wayne Hoffman, *The End of Her* (Heliotrope Books, 2022)
- Menachem Kaiser, *Plunder: A Memoir of Family Property and Nazi Treasure* (Houghton Mifflin Harcourt, 2021)
- Erika Krouse, *Tell Me Everything* (Flatiron Books, 2022)
- Treva B. Lindsey, *America, Goddam* (University of California Press, 2022)
- Rachel Rear, *Catch the Sparrow* (Bloomsbury, 2022)

- Liza Rodman and Jennifer Jordan, *The Babysitter: My Summers with a Serial Killer* (Atria, 2021)
- Jacqueline Rose, *On Violence and On Violence Against Women* (Farrar, Straus & Giroux, 2021)
- Javier Sinay, *The Murders of Moisés Ville* (Restless Books, 2022)
- Emma Southon, *A Fatal Thing Happened on the Way to the Forum* (Abrams, 2021)
- Tori Telfer, *Confident Women* (Harper Paperbacks, 2021)
- Natasha Trethewey, *Memorial Drive* (Ecco, 2020)

PODCASTS

- *9/12* (Amazon Originals, 2021)
- *Believe Her* (Lemonada Media, 2021)
- *Bone Valley* (Lava For Good, 2022)
- *Canary* (The *Washington Post* Investigates, 2020)
- *Crime Show* (Gimlet Media, 2021–2022)
- *Crimes of the Centuries* (Obsessed Network, continuing)
- *Dead End* (WNYC Studios, 2022)
- *Document: Death Resulting* (New Hampshire Public Radio, 2021)
- *Do You Know Mordechai?* (UCP Audio, 2021)
- *The Line* (Apple Original/Jigsaw, 2021)
- *Love Is a Crime* (C13 Originals/*Vanity Fair,* 2021)
- *Love Thy Neighbor* (Pineapple Street Studios, 2022)
- *Mississippi Goddam* (Reveal/CIR, 2021)
- *Murderville, Texas* (The Intercept, 2022)
- *Project Unabom* (Apple Original/Pineapple Street Studios, 2022)

- *Stolen: The Search for Jermain* and *Stolen: Surviving St. Michael's* (Gimlet Media, 2021–2022)
- *Suave* (Futuro Media Group/PRX, 2021)
- *Suspect* (Wondery/Campside Media, 2021–2022)
- *Through The Cracks* (WAMU, 2021)
- *Undisclosed* (self-produced, ended in 2022)
- *What Happened to Sandy Beal* (iHeart, 2022)
- *Wind of Change* (Pineapple Street Studios/Crooked Media/Spotify, 2020)

FILM AND TELEVISION

- *Atlanta's Missing and Murdered: The Lost Children* (HBO, 2020)
- *Captive Audience* (Hulu, 2022)
- *The Case Against Adnan Syed* (HBO, 2020)
- *The Dropout* (Hulu, 2022)
- *Fear City: New York vs. the Mafia* (Netflix, 2020)
- *Girl in the Picture* (Netflix, 2022)
- *How to Fix a Drug Scandal* (Netflix, 2020)
- *I'll Be Gone in the Dark* (HBO, 2020)
- *The Innocence Files* (Netflix, 2020)
- *Lost Girls* (Netflix, 2020)
- *Only Murders in the Building* (Hulu, 2021–2022)
- *The Tinder Swindler* (Netflix, 2022)
- *Trial by Media* (Netflix, 2020)
- *Under the Banner of Heaven* (FX, 2022)
- *Unsolved Mysteries* (Netflix, ongoing)
- *We Need to Talk About Cosby* (Showtime, 2022)

NEWSLETTERS

Really, there is only one of substance: *Best Evidence,* helmed by Sarah D. Bunting and Eve Batey, which covers true crime film, television, podcasts, and books each weekday with careful critical attention, analysis, scrutiny, and wry humor. I'm a faithful subscriber. It's available at bestevidence.substack.com.

CONTRIBUTORS

LARA BAZELON is a professor at the University of San Francisco School of Law, where she holds the Barnett Chair in Trial Advocacy and directs the criminal and racial justice clinics. She is the author of three books, most recently *Ambitious Like a Mother: Why Prioritizing Your Career Is Good for Your Kids* (Little, Brown, 2022). Her op-eds, essays, and long-form journalism pieces about crime, justice, love, work, and family have been published in the *New York Times,* the *Washington Post,* the *Atlantic,* and *New York Magazine,* among other media outlets. She lives in San Francisco with her two children.

KERI BLAKINGER is the author of *Corrections in Ink,* a memoir about her time in prison. She now works as an investigative reporter for the Marshall Project and previously covered criminal justice for the *Houston Chronicle.* Her work has appeared in the *Washington Post Magazine, Vice,* the *New York Daily News,* and the *New York Times.*

RABIA CHAUDRY is an attorney, author, advocate, and the executive producer of the four-part HBO documentary *The Case Against Adnan Syed,* based on her *New York Times* best-selling book, *Adnan's Story*. Her critically acclaimed second book, *Fatty Fatty Boom Boom: A Memoir of Food, Fat, and Family,* was published by Algonquin in November 2022. Rabia is cohost and coproducer of five podcasts, including *Undisclosed,* the most popular wrongful conviction podcast in the world. A 2021 Aspen Institute/ADL Civil Society Fellow and a 2016 Aspen Ideas Scholar, she serves on the Vanguard Board at the Aspen Institute. She is a fellow of the Truman National Security Project, the American Muslim Civic Leadership Institute, and the Shalom Hartman Institute, and a founding board member of the Inter Jewish Muslim Alliance and the Muslim-Jewish Advisory Council, both of which focus on building Muslim-Jewish coalitions around pressing policy issues and educating across communities to break barriers. To learn more about Rabia, please visit rabiachaudry.com.

SOPHIE HAIGNEY is the web editor at the *Paris Review*. She writes about books and culture for the *New York Times, The New Yorker,* and other publications. She is currently at work on an essay collection about objects and collecting.

MICHAEL HOBBES is the cohost of the podcast *Maintenance Phase* and the founding cohost of *You're Wrong About*. His work has appeared in *Slate, HuffPost, Foreign Policy,* and *Pacific Standard.*

MAY JEONG is a writer for *Vanity Fair*. Her reporting from Afghanistan has been awarded the South Asian Journalist Association's Daniel Pearl Award, and the Bayeux Calvados Normandy Award for War Correspondents, and has also been recognized by the Kurt

Schork and Livingston Awards. She was awarded the 2022 J. Anthony Lukas Work-in-Progress Award and the Whiting Creative Nonfiction Grant for her upcoming book on sex work. She lives on land ceded by the Lenape people in the Treaty of Shackamaxon in 1682, also known as Brooklyn, New York.

RF JURJEVICS works in tech by day and does just about everything else possible by night. Their work has been published in *Real Simple, Vice,* the *San Diego Reader, Slate,* Dame, GOOD, and *Allure*.

AMANDA KNOX is an exoneree, journalist, public speaker, author of the *New York Times* best-selling memoir *Waiting to Be Heard,* and cohost, with her partner Christopher Robinson, of the podcast *Labyrinths*. Between 2007 and 2015, she spent nearly four years in an Italian prison and eight years on trial for a murder she didn't commit. She has since become an advocate for criminal justice reform and media ethics. She sits on the board of the Frederick Douglass Project for Justice.

JUSTINE VAN DER LEUN is an author and independent journalist. Her books include, most recently, *We Are Not Such Things,* and she is the host, lead reporter, and coproducer of the award-winning investigative podcast *Believe Her*. Justine's reporting has been recognized or supported by the Columbia University School of Journalism, the Emerson Collective, New America, Type Media Center, the Pulitzer Center, the Society of Professional Journalists, and PEN America, among others.

WESLEY LOWERY is a Pulitzer Prize–winning journalist and author specializing in issues of race and law enforcement. He led the

Washington Post team awarded the Pulitzer Prize for National Reporting in 2016 for the creation and analysis of a real-time database to track fatal police shootings in the United States. His 2018 project, "Murder with Impunity," an unprecedented look at unsolved homicides in major American cities, was a finalist for the Pulitzer Prize. His first book, *They Can't Kill Us All: Ferguson, Baltimore, and a New Era in America's Racial Justice Movement,* was a *New York Times* bestseller and was awarded the Christopher Isherwood Prize for Autobiographical Prose by the LA Times Book Prizes.

BRANDI MORIN is an award-winning Cree/Iroquois/French journalist from Treaty 6 territory in Alberta and the author of *Our Voice of Fire: A Memoir of a Warrior Rising*. Her work has appeared in publications and networks including *National Geographic,* Al Jazeera English, the *Guardian, Vice, ELLE Canada,* the *Toronto Star,* the *New York Times, Canadaland, HuffPost,* Indian Country Today Media Network, the Aboriginal Peoples Television Network National News, and CBC Indigenous.

DIANA MOSKOVITZ is a cofounder, co-owner, and investigations editor at Defector Media. She has been recognized for her reporting by the Investigative Reporters & Editors Awards, the Deadline Club Awards, *Longreads, Longform,* and more. Her work also was featured in *The Year's Best Sports Writing 2022*. Originally from South Florida, Diana lives in Los Angeles, where she is currently working on her first book.

MALLIKA RAO is an award-winning essayist and reporter, and a new writer of fiction. Her byline can be found or is forthcoming at *New York* magazine, *Harper's,* and the *Believer*. She's presently at work on a collection of short stories. Her anthologized essay, "Three Bodies in Texas," is currently being developed into a podcast.

AMELIA SCHONBEK is a journalist whose work often focuses on how people experience and recover from trauma. She contributes regularly to *New York* magazine. When not at work, she likes to sing, swim, and be outside.

SAMANTHA SCHUYLER is a writer and fact-checker from Florida living in New York City.

PERMISSIONS